HS●AC
HOMELAND SECURITY
OPERATIONAL ANALYSIS CENTER

Development of Standardized and Best Practices for the USCG Boats Acquisition Program

BRENDAN TOLAND, MICHAEL VASSEUR, AARON C. DAVENPORT, SCOTT SAVITZ, KATHERYN GIGLIO

Published in 2019

Preface

The U.S. Coast Guard (USCG) depends on its fleet of more than 1,600 boats to conduct its most critical operations, which span all 11 of the USCG's statutory missions.[1] These boats must be replaced frequently, given the harsh environments and challenging operations in which they are used.

To keep up with this demand in a cost-effective way, the USCG has determined that it needs an enduring Program Management Office to manage boats acquisition efforts. The RAND Corporation's Homeland Security Operational Analysis Center (HSOAC) was asked to conduct a 90-day study to (1) identify best practices and lessons learned for improving boats acquisition by reviewing the current boats acquisition program and similar programs inside and outside the USCG and (2) make recommendations for the structure, funding strategy, and processes of a future enduring boats acquisition program. We review the current boats acquisition program and similar organizations inside and outside the USCG, assess possible funding and structural strategies, and make recommendations on these topics for USCG leadership.

This report should be of interest to USCG stakeholders and those concerned with watercraft acquisition across the Departments of Defense and Homeland Security.

This research was sponsored by the USCG and conducted within the Acquisition and Development Program of the HSOAC federally funded research and development center (FFRDC).

About the Homeland Security Operational Analysis Center

The Homeland Security Act of 2002 (Section 305 of Public Law 107-296, as codified at 6 U.S.C. § 185), authorizes the Secretary of Homeland Security, acting through the Under Secretary for Science and Technology, to establish one or more FFRDCs to provide independent analysis of homeland security issues. The RAND Corporation operates the Homeland Security Operational Analysis Center (HSOAC) as an FFRDC for the U.S. Department of Homeland Security (DHS) under contract HSHQDC-16-D-00007.

The HSOAC FFRDC provides the government with independent and objective analyses and advice in core areas important to the department in support of policy development, decisionmaking, alternative approaches, and new ideas on issues of significance. The HSOAC FFRDC also works with and supports other federal, state, local, tribal, and public- and private-sector organizations that make up the homeland security enterprise. The HSOAC FFRDC's

---

[1] The 11 missions are (1) search and rescue; (2) marine safety; (3) ports, waterways, and coastal security; (4) defense readiness; (5) migrant interdiction; (6) drug interdiction; (7) aids to navigation and waterway management; (8) ice operations; (9) living marine resources; (10) other law enforcement; and (11) marine environmental protection.

research is undertaken by mutual consent with DHS and is organized as a set of discrete tasks. This report presents the results of research and analysis conducted under HSCG23-17-J-ADW042.

The results presented in this report do not necessarily reflect official DHS opinion or policy. For more information on HSOAC, see www.rand.org/hsoac.

For more information on this publication, visit www.rand.org/t/RR2918.

Contents

Figures

Tables

Summary

The U.S. Coast Guard (USCG) depends on boats—watercraft less than 65 feet in length and operated from the shoreline and from cutters—to conduct critical operations that span all of the USCG's 11 statutory missions. The USCG currently has an inventory of more than 1,600 active boats that, given the harsh environments and challenging operations in which they are used, eventually become too expensive to maintain, often after only a decade or less of service. As a result, the USCG is continually in the process of acquiring new boats for cutters and shore-based stations.

To keep up with this demand in the most cost-effective way, the USCG has determined that it needs an enduring program management office (PMO)—the USCG Boats Acquisition Program Office (CG-9325)—to manage boats acquisition efforts. The RAND Corporation's Homeland Security Operational Analysis Center (HSOAC) was asked to assist in this effort in two ways: (1) Identify best practices and lessons learned for improving boats acquisition and (2) make recommendations for the structure, funding strategy, and processes of an enduring boats acquisition program. We reviewed the current boats acquisition program and similar organizations inside and outside the USCG. In addition, we reviewed key policy documentation and plans from the Department of Homeland Security (DHS), the USCG, and other organizations with similar acquisition needs, such as the U.S. Navy. Researchers also conducted more than 25 interviews with members of the USCG boats program, stakeholders in the USCG, and other related organizations.

We reviewed and compared seven organizations—three inside the USCG and four outside—that fulfill acquisition functions similar to those the USCG needs for its boats acquisition program. Specifically, we examined how these programs are structured, funded, and staffed. We developed and compared different structure and funding options, based on our review of these internal and external organizations. We assessed these options according to factors deemed important to the USCG's new boats acquisition program through a review of USCG boats acquisition history, policy, and interviews. Four factors served as metrics for comparisons between options:

1. **Ability to add additional projects**: An enduring program will likely be needed to manage unforeseen boats acquisition. An effective program structure and funding strategy will minimize the challenge of adding new acquisition projects.
2. **Flexibility:** Boats acquisitions needs can change, even once a program is underway. An effective program structure and funding strategy will be flexible.
3. **Ability to interface with other stakeholders:** During the course of acquisition, the PMO will need to work with a variety of other USCG offices to ensure the boats acquired meet the needs of all parts of the organization.
4. **Staffing:** We took into consideration the number of staff; the distribution of military, civilian, and contractor staff; and the relative composition of program management,

logistics, and technological skills within any potential PMO. We also considered the stability of staffing over time and general staffing demands associated with particular program structures and funding strategies.

Summary of Recommendations

To meet the challenges described above, we offer 11 recommendations, summarized in Figure S.1. The recommendations are organized into three broad categories: policy recommendations, long-term (or strategic) recommendations, and near-term (or tactical) recommendations. In addressing the primary study issues—program structure and the acquisition process—the researchers uncovered a number of additional issues and potential recommendations, which are included in the table. (Chapter 5 in the full report discusses all recommendations in greater detail, providing insights from stakeholder interviews, observations, and analytic context.)

Figure S.1. Summary of Recommendations

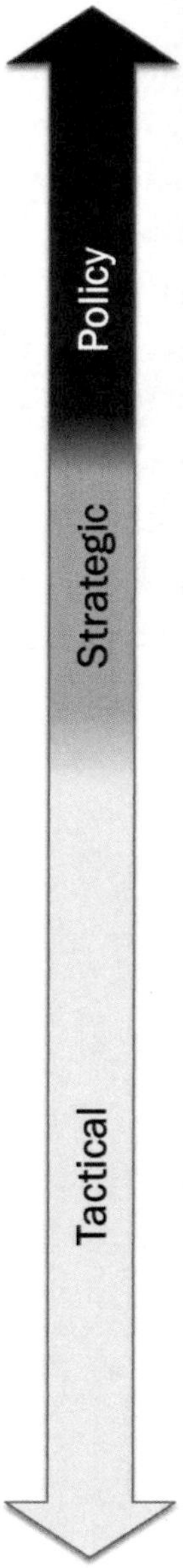

Focus Area	Recommendation
Guidance Ambiguity	Resolve ambiguity introduced by PL-113-6 allowing for Operating Expense (OE) expenditures for "contingent and emergent requirements" with current practices
Below-NMAP Guidance	Establish guidance for the prescribed acquisition process that should be followed for purchases of boats below the non-major ($10M) threshold
Capital Asset Management Plan	Develop a Boat Capital Asset Management Plan analogous to the Cutter Capital Asset Management Plan (CCAMP)
Goals and Metrics	Continue to develop program goals from which verifiable metrics can be derived that are traceable to CG-9 guidance
Standardization of Boats	Continue efforts to standardize boat fleet and reduce risks and costs associated with acquisition and maintenance
Clarity on Roles	Designate one office as the sole procurer of boats for the USCG to provide increased efficiencies and decreased uncertainty among stakeholders
Funding Strategy	Utilize a mixed funding strategy using both OE and AC&I with an emphasis on an AC&I direct appropriation to the boats acquisition program office
Structure Strategy	Utilize an augmented matrix structure based on boat type vice individual boat projects
Staffing Strategy	Conduct a rigorous analysis of anticipated workloads for the boats program and staff. Other comparable organizations inside and outside the USCG have staff sizes between 20-30 personnel.
Acquisition Analysis	Conduct predictive analysis on inventory, in coordination with CG-731, to better prepare for new projects
Collaboration	Explore possible collaboration with boat acquisition offices external to USCG

The vertical axis (top to bottom) spans from Policy / Strategic at the top to Tactical at the bottom.

NOTE: OE = Operating Expense; AC&I = Acquisition, Construction, and Improvement.

The policy-related recommendations at the top of Figure S.1 relate to high-level guidance at the DHS or USCG level and are focused on resolving ambiguity and the lack of guidance in certain aspects of boats acquisition. The *Financial Resource Management Manual* (FRMM) provides clear guidance on funding boats acquisition, but the flexibility provided by Public Law 113-6 introduces some uncertainty. There is also ambiguity in the procurement process for systems falling below the nonmajor ($10 million) threshold. Although these challenges may require solutions external to the USCG Office of Surface Acquisition (CG-932) that are beyond the scope of this study, we note that development of a document similar to the *Cutter Capital*

Asset Management Plan—but specific to boats—could aid in detailing program scope, roles and responsibilities, inputs and outputs, and definitions of terms.

The strategic, long-term recommendations in the center of Table S.1 focus on achieving major program goals of CG-9325. They are amplifications of initiatives already underway within the boats program. We recommend that CG-9325 continue to develop program goals and metrics that are traceable to USCG Acquisition Directorate (CG-9) guidance and continue to examine methods to introduce more standardization into the boat fleet. Our review of inventory data indicated that the USCG has more than 1,600 boats in its inventory, with 49 to 72 distinct models. Some reduction in the number of boat models acquired and the number of boat manufacturers would likely reduce risks and costs associated with acquisition, operation, maintenance, and logistics.

We provide several near-term, tactical recommendations for CG-9325 to improve operations within the program at the bottom of Figure S.1. These include core issues of funding strategy, staffing strategy, and organizational structure. In addition, we call for future acquisition analysis and possible collaboration with external organizations. Finally, we recommend designating one organization as the procurer of boats for the USCG to continue to reduce ambiguity.

Acknowledgments

We thank the U.S. Coast Guard Office of Surface Acquisition for providing us with guidance and support during the study and particularly for providing access to relevant experts and written sources. We greatly appreciate the efforts of Glen Maylone, LCDR Christopher Lavin, Mark Porvaznik, and CDR David Obermeier.

We are also greatly appreciative of the time and information provided by our interviewees from the following offices: Boats Acquisition (CG-9325); Human Systems Integration (CG-1B3); Office of Boat Forces (CG-731); Office of Budget and Programs (CG-82); Office of Naval Engineering (CG-45); Small Boat Product Line, C4ISR Program (CG-9335); Electronic Health Record Acquisition, In-Service Vessel Sustainment (CG-9323); Army Corps of Engineers Marine Design Center; Customs and Border Protection Air and Marine; Naval Sea Systems Command, Support Ships, Boats, and Craft Program Office; and the National Oceanic and Atmospheric Administration Offices of Law Enforcement and Office of Fisheries.

We also thank RAND Corporation colleagues who provided valuable comments and inputs to the study, specifically Isaac Porche, Erin-Elizabeth Johnson, and CAPT John Rivers. Jeffery Drezner and Thomas Atkin provided peer reviews and as a result greatly improved earlier versions of this report.

Abbreviations

AC&I	Acquisition, Construction, and Improvement
APM	assistant program manager
ATON	Aid to Navigation
C4ISR	Command, Control, Communications, Computers, Intelligence, Surveillance, and Reconnaissance
CBP	Customs and Border Protection
CCAMP	Cutter Capital Asset Management Plan
CG-731	USCG Office of Boat Forces
CG-751	USCG Office of Cutter Forces
CG-82	USCG Office of Budget and Programs
CG-9	USCG Acquisition Directorate
CG-932	USCG Office of Surface Acquisition
CG-9323	USCG In-Service Vessel Sustainment Program
CG-9325	USCG Boats Acquisition Program
CG-9335	USCG C4ISR Program
DHS	Department of Homeland Security
EHRA	Electronic Health Records Acquisition
FRMM	Financial Resource Management Manual
HSOAC	Homeland Security Operational Analysis Center
ISVS	In-Service Vessel Sustainment
NAVSEA	Naval Sea Systems Command
NAVSEA PMS-325	Naval Sea Systems Command, Support Ships, Boats, and Craft
NOAA	National Oceanic and Atmospheric Administration
OE	Operating Expense
PM	project manager
PMO	Program Management Office
SLEP	Service Life Extension Program
SME	subject-matter expert
USCG	U.S. Coast Guard

1. Introduction

Boat-based operations are central to U.S. Coast Guard (USCG) missions. As of February 2017, the USCG had an inventory of more than 1,600 active boats—watercraft less than 65 feet in length. Boats are operated both from the shoreline and from cutters (much larger, typically habitable watercraft).[1] In many cases, boat operations are the culmination of a mission: Boats are used to conduct boardings, interdictions, and rescues; deter and counter terrorist attacks; and maintain aids to navigation. It would be difficult to overstate their importance to overall USCG effectiveness.

USCG boats must be capable, reliable, and safe in diverse, challenging environments. The USCG often has limited choices about the circumstances in which it can deploy its boats, since many missions are time sensitive. USCG boats need to be durable and stable to withstand rough seas. They need to allow safe performance of such duties as recovering people, deploying or retrieving boarding parties, and using weapons. Their command, control, communications, computers, intelligence, surveillance, and reconnaissance (C4ISR) systems must be fit to be used in harsh conditions, such as salt spray or cold weather, by crew members wearing gloves. Boats need to have a level of comfort so that personnel can perform missions in the short term and avoid chronic conditions in the long term. Finally, boats need to be able to last long enough in harsh maritime environments to be cost effective, and they need to be designed so that they can be maintained for capable performance. (Most boats become too expensive to maintain after a decade or less of service, given the harsh environments and challenging operations in which they are used).

Given these stringent needs and the USCG's high demand for large numbers of boats, a well-managed boats acquisition process is critical.

The USCG replaces boats based on the following factors: a boat's operational availability, cost per operating hour, and ability to successfully complete assigned missions. If any one of these three factors is significantly affected, a decision to replace will be considered.[2] New boats are acquired either through Operating Expense (OE)– or Acquisition, Construction, and Improvement (AC&I)–funded contracts. Acquiring new boats requires specific documentation prior to funding, to include an Operational Requirements Document, Capability Analysis Report, Mission Needs Statement, and Concept of Operations. The Office of Boat Forces (CG-731) prepares this documentation in close consultation with field units, technical authorities, subject-

---

[1] There is some ambiguity about the number of boats, since sources vary. Discrepancies appear to reflect, at least partly, how boats are counted (e.g., boats that are coming into or going out of the inventory might be counted by some sources but not others), as well as latent reporting of acquisition or disposal. Totals provided range from 1,400 to 1,800 boats.

[2] Commandant Instruction Manual 16114.4B, *Boat Management Manual*, Washington, D.C.: USCG, 2012.

matter experts (SMEs), and other program managers, support activities, and stakeholders. Nonmajor acquisitions have historically been purchased using OE funding.[3] Lifecycle replacement and repair of supporting equipment is the responsibility of the Small Boat Product Line at the Surface Forces Logistics Center and the operating field unit through recurring Standard Support Level funding. Only CG-731 may authorize boat purchases.

The USCG has determined that it needs an enduring program management office (PMO) to manage those efforts. The current PMO, USCG's Boats Acquisition Program (CG-9325), is the remnants of the former PMO for the Response Boat–Medium acquisition. This program has delivered all its assets and is closing out. CG-9325 is now chartered as the Response Boat Small-II project manager (PM), Motor Life Boat Service Life Extension Program (SLEP) PM, as well as the Cutter Boats PM. The Response Boat–Medium and most new acquisition programs are funded with AC&I funds; however, CG-731 also purchases boats using OE funding. These acquisitions vary across DHS acquisition levels and include major and nonmajor acquisition programs. In some instances, CG-9325 and CG-731 use the same contracts with different funds to purchase the same class of boat for different host units. CG-9325 provides program management support to CG-731 in some of these efforts.

Study Purpose

The USCG asked the RAND Corporation's Homeland Security Operational Analysis Center (HSOAC) to provide analysis and recommendations for how best to structure and fund a future USCG office, including both assets and personnel, to acquire boats in a resource-constrained environment. Specifically, this study has two purposes: (1) identifying best practices and lessons learned for improving boats acquisition by reviewing the current boats acquisition program (e.g., CG-9325) and similar organizations inside and outside the USCG and (2) making recommendations for the structure, funding strategy, and processes of a future, enduring boats acquisition program. Figure 1.1 presents the study team's four primary tasks.

---

[3] *Nonmajor acquisitions* are defined as procurements greater than $10 million in procurement costs and less than $300 million in lifecycle costs that are not designated as a major systems acquisition (Commandant Instruction Manual 5000.11A, *Non-Major Acquisition Process [NMAP] Manual*, Washington, D.C., 2011).

Figure 1.1. The Four Steps of the USCG Boats Acquisition Study

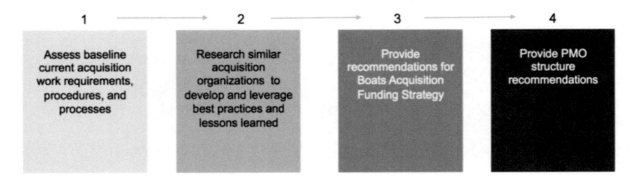

This report provides an independent review of the current small boats program and identifies best practices in other organizations that it can leverage. As a federally funded research and development center, HSOAC is uniquely positioned to provide conflict-free recommendations. This report will be valuable to the USCG acquisition community for future program planning and organization.

Description of Study Methods and Limitations

To complete the study tasks, HSOAC conducted research through two primary means: document review and interviews. We reviewed DHS, USCG, and CG-9 guidance and documentation.[4] Additionally, CG-9325 provided several draft and working-level documents concerning small boats management and acquisition. These include a Boat Program inventory, descriptions of the enterprise landscape, vision and functional statements, and the Plan of Actions and Milestones.

HSOAC conducted more than 25 interviews with members of the Boats Program, stakeholders within the USCG, and other acquisition organizations both inside and outside of the USCG that manage acquisition programs facing similar challenges. The interviews allowed the researchers to better understand the perspectives and opinions of the individuals working in these

---

[4] These include the following primary authoritative acquisition management documents:

- Commandant Instruction Manual 5000.10D, *Major Systems Acquisition Manual (MSAM)*, Washington, D.C.: USCG, 2015
- Commandant Instruction Manual 5000.11A, 2011
- Commandant Instruction Manual 7100-3E, *Financial Resource Management Manual (FRMM)*, Washington, D.C.: USCG, 2013
- DHS, Under Secretary for Management, *Acquisition Management Instruction*, Washington, D.C.: USCG, DHS Instruction 102-01-001-R01, March 9, 2016.
- USCG, Acquisition Directorate, *Strategic Plan: Blueprint for Sustained Excellence*, Version 6.1, Summer 2016.
- Commandant Instruction Manual 5000.12, *Management Roles and Responsibilities*, Washington, D.C.: USCG, 2012.

offices and discuss their issues, challenges, and recommendations for improvement in boats acquisition. Individuals from the following organizations were interviewed:

- CG-9325
- Relevant USCG stakeholders and partners

    - Human Systems Integration (CG-1B3)
    - Office of Boat Forces (CG-731)
    - Office of Cutter Forces (CG-751)
    - Office of Budget and Programs (CG-82)
    - Office of Naval Engineering (CG-45)
    - Small Boats Product Line

- Potentially analogous USCG offices

    - C4ISR Program (CG-9335)
    - Electronic Health Record Acquisition (EHRA)
    - In-Service Vessel Sustainment (ISVS; CG-9323)

- Organizations external to USCG

    - Army Corps of Engineers Marine Design Center
    - Customs and Border Protection (CBP) Air and Marine
    - Naval Sea Systems Command, Support Ships, Boats, and Craft (NAVSEA PMS-325)
    - National Oceanic and Atmospheric Administration (NOAA) Offices of Law Enforcement and Office of Fisheries.

Study Limitations

While assessing the conclusions drawn in this report, it is important to keep several limitations in mind. First, this report is based on a 90-day study; the study's quick-turn nature limited the depth and breadth of the analysis. Second, many of the processes examined in this report are currently in flux among all organizations under DHS oversight, as DHS refines its own processes. Third, we assessed a limited number of organizations to draw lessons learned for CG-9325. We selected organizations that either acquire boats or face similar acquisition challenges within the USCG, but not an exhaustive list of all possible organizations. We did not assess the analogous organizations themselves but instead examined how they operate today in ways that might be useful to the USCG boats acquisition program. The best practices we identify from these examinations are those that SMEs from each organization interviewed identified as being key to their acquisition processes. Given the timing and scope of this study, we did not independently verify the efficacy of these practices. Finally, many acquisitions occurred in a time before the current organizational structure, making it impossible to obtain detailed data on these projects or consider these data in our analyses.

Organization of This Report

Chapter 2 describes the current state of boats acquisition in the USCG. Chapter 3 addresses how other organizations, both internal and external to the USCG, address similar acquisition challenges. Chapter 4 provides an assessment of possible funding strategies and organizational structures for a boats acquisition PMO. Chapter 5 summarizes our findings and offers recommendations for USCG leadership to consider going forward.

2. Current Small Boats Acquisition Work Requirements, Procedures, and Processes

This chapter presents a full description of the current state of boats acquisition in the USCG. The findings in this chapter are critical, as they both define the need for a separately designated USCG boats acquisition program and provide the baseline by which we can begin to assess appropriate strategies for the development of such a program. The chapter begins by describing the acquisition challenge of acquiring boats for the USCG by focusing on the two offices with central roles in the current boats acquisition system. The chapter closes with an assessment of key areas of the current process. The information and assessment in this chapter inform the baseline case of USCG work requirements, procedures, and processes used throughout this study.

USCG Acquisition Overview

The acquisition process that is used to purchase boats under AC&I appropriations is defined in DHS and Coast Guard directives and policy guidance. The PMO is responsible for planning and executing an acquisition project within an established cost, schedule, and performance baseline. The acquisition PMO performs the following functions:

1. plans, organizes, executes, and coordinates the assigned acquisition project in accordance with approved and applicable acquisition policies, processes, and procedures[1]
2. executes the core processes and activities with participation from appropriate stakeholders, including sponsor(s), legal, and technical authorities
3. acts as the focal point for reporting project specific information as required, in accordance with acquisition directives
4. serves as the principal adviser to all formal project specific source-selection activities
5. assists sponsor with requirements generation activities.

The challenge is in providing an enduring program governance model and staffing infrastructure to efficiently and effectively acquire boats to support the service's missions and statutory mandates, maximizing the return on investment. Figure 2.1 shows the program manager's role as integrator of the three primary areas of management into a coherent strategy to achieve cost, schedule, and performance objectives for their program.

---

[1] Commandant Instruction Manual 5000.12, 2012; Major Systems Acquisition Manual, Commandant Instruction Manual 5000.10D, 2015; Commandant Instruction Manual 5000.11B, *Non-Major Acquisition Process (NMAP) Manual*, Washington, D.C.: U.S. Coast Guard, 2012.

Figure 2.1. Management Interfaces

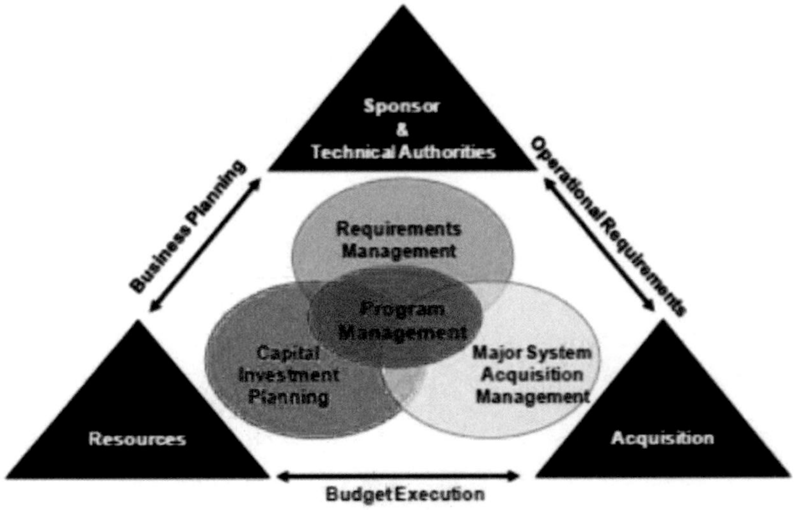

SOURCES: Commandant Instruction Manual 5000.12, 2012; Commandant Instruction Manual 5000.10D, 2015.

USCG boats are procured using commercial boats, commercial boats with modifications, and clean sheet designs. Most frequently, they are procured from parent designs or commercially available boats with modifications. Ideally, what is needed is a deliberate, single repository for design characteristics data for the entire boats inventory. This does not exist and is not something that can be readily obtained from records. The Federal Acquisition Regulation generally requires purchase order records be retained for only four years, so records disappear long before most of these boats reach the end of service life.

However, regardless of a particular boat's design, the USCG's acquisition strategy is the same. The process is driven by both an individual and total program cost threshold. The less deviation from the original parent design, the cheaper and faster the delivery can be. The majority of modifications involve either government-furnished equipment or commercial off-the-shelf equipment, which is added on or after delivery or through a subcontractor who works with the prime contractor to make modifications during the build process. Ships are commonly delivered along with their boats, as the small boat is regarded as an integral portion of a ship's overall mission capability.

Two Offices Share Responsibility for USCG Boats Procurement

The USCG acquires a significant number of boats each year. Their characteristics make them challenging acquisitions. Most boats have a short lifecycle; on average, they last less than ten years. This requires frequent recapitalization or service life-extension projects. This is further complicated by the existence of more than 50 different boat types. The current process of procuring boats for the USCG is split between two offices: CG-9325 and CG-731. In this section, we describe the roles each office plays in the USCG acquisition of boats under the current system generally and as it relates to key sections of the acquisition process.

CG-9325

CG-9325 is responsible for acquiring the service's boat fleet and manages the acquisition of boats for projects that move to the acquisition needs phase (called Acquisition Decision Event One). CG-9325's mission is to efficiently acquire and deliver standard classes of multimission boats on budget that meet or exceed operational threshold requirements by managing multiple appropriations through a disciplined acquisition process. Moreover, the boats acquisition program seeks to provide strategic acquisition planning and execution, which enables the sponsor to focus on strategic capability planning, conducting field unit outreach, and maintaining external partnerships throughout the acquisition process. CG-9325 seeks to deliver value through standard practices that consistently integrate the sponsor, sustainment community, and other key stakeholders in all phases of the acquisition process, with the ultimate goal of achieving excellence in system delivery, field introduction, initial training, and logistics support.

CG-9325 Responsibilities and Management

Considerable planning is required to fully execute boats acquisition, given the number of assets and their short service life. Currently, CG-9325 manages six different boats acquisition projects. These six projects, in turn, are divided into four major groupings:

- **Cutter boats:** Cutter boats are generally between 18 and 34 feet and are carried aboard USCG cutters. They can be deployed for such missions as surveillance and interdiction. There are 19 different types of cutter boats, of which ten types have been classified as beyond their useful life.
- **Response boats:** Response boats conduct similar missions to cutter boats, but are launched from shore-based USCG stations. There are 22 types, of which seven types are beyond their useful life.
- **Aid-to-navigation (ATON) boats:** ATON boats are used to maintain buoys, ranges, and other fixed aids to navigation, and can reach up to 64 feet in length. This group comprises seven types, with three types beyond their useful life.
- **Training boats:** These boats are used to train future USCG boat operators. There are 12 types in this group; three types have been assessed as beyond their useful life.[2]

Useful life is defined as the point in an asset's lifecycle when maintenance availability exceeds service standards, operational availability is sufficiently degraded, and the actual support costs exceed the standard support level that has been established within the existing budget model. Program data indicates that, at the time of writing, over one-third of the small boat fleet is in need of a planned replacement project.

---

[2] Information on boat types and life stage is drawn from data provided by CG-731 and CG-9325 as well as context provided in SME interviews.

CG-9325 uses an organizational matrix-based approach to acquire boats, assigning an assistant program manager (APM) to manage a project across the acquisition process. Additional personnel within the PMO are organized into functionally based groups and are tasked to specific projects as required.[3] The current PMO is staffed by a mix of approximately 15 military, civilian, and contractor personnel directly assigned to CG-9325. CG-9325 also relies heavily on other headquarters PMOs and field support activities to complete all the actions needed throughout the acquisition process that lead to asset delivery and handoff. Unlike many other PMOs, CG-9325 does not have on-site project residence offices with manufacturers to deal with quality concerns and coordinate deliveries with the field. This work is instead undertaken by PMO staff located at USCG headquarters.

CG-9325 Funding and Acquisition Processes

Boats are acquired through AC&I and OE funds. AC&I funds are five-year funds that cover the acquisition or improvement of major USCG systems including vessels, aircrafts, and shore installations. OE funds are one-year funds that pay for general USCG operating expenses in a variety of categories. The FRMM provides clear guidance on when to use OE or AC&I funds, stating that OE is to be used for "unit-level recurring maintenance and repair" and "minor repair and corrective maintenance."[4] By this definition, all boats should be acquired via AC&I funds. However, the USCG has received a legislative exception to this rule that allows for the purchase of boat projects using OE funds, provided that individual assets cost less than $700,000, no more than $31 million is spent on boats in a given fiscal year, and they support "contingent and emergent requirements."[5] Currently, if the cost is under a $700,000 threshold, boats are typically purchased using OE funds.

CG-9325 receives AC&I funds to purchase boats above the $700,000 threshold. Some of these funds come through a direct appropriation related to the purchasing of cutter boats for the National Security Cutter. Other cutter programs also supply AC&I funds to CG-9325 to place against their contracts for the purchase of cutter boats, but these funds are not directly controlled by CG-9325.

CG-9325 focuses on acquiring boats that are approved as major or nonmajor acquisitions. As these systems represent a significant investment for the USCG, they require approval authority outside of CG-9325. In the case of major systems, the approval authority rests with DHS.

---

[3] See Harold Kerzner, *Project Management: A Systems Approach to Planning, Scheduling and Controlling*, 8th ed., Hoboken, N.J.: John Wiley & Sons, Inc., 2003, for further details.

[4] Commandant Instruction Manual 7100-3E, 2013.

[5] Pub. L. 113-6, Consolidated and Further Continuing Appropriations Act, March 26, 2013.

CG-731

CG-731 is the lead policy office and sponsor for the USCG boat program. CG-731 is charged with providing oversight to support safe and effective boat operations, building and sustaining the right capability to meet mission requirements, and establishing requirements for the management of the service fleet. CG-731 is the sole office that may authorize boat purchases.[6] CG-731 is also responsible for funding boat replacements, and in some cases executes boat procurements within the office.

CG-731 Responsibilities and Management

The acquisition of new boats requires appropriate documentation. CG-731 derives this documentation with field unit input and works in coordination with headquarters program managers during each phase of the procurement process. To continue this process on a recurring basis, CG-731 requires the regional boat managers to provide an annual report that communicates both cutter and shore-based boat replacement needs over the next two fiscal years. The annual reports are used to document, justify, and prioritize replacement to inform the budgeting process.

The CG-731 boat replacement philosophy is driven by business-case decisions. Replacement occurs when it makes sense from a business case perspective to do so—a determination driven by assessments of the operational availability rate, cost per operating hour, and the ability of the asset to meet its assigned missions. Replacement need is prioritized based on both qualitative and quantitative criteria, including safety, age, materiel condition, cost to maintain, and estimates of needed repairs to extend service life. CG-731 prepares a host of justifications through the preparation and collection of documents and performs administrative actions to further document need and individual asset requirements.

In addition to the *Boat Management Manual*, boats acquisition is guided by the *Shore Based Response Boat Strategic Vision and Transition Plan*, which mandates and underscores the need for standardized boats, presents the service's plan to standardize wherever possible in order to enhance efficiencies across the small boat support enterprise, and seeks to maximize operational availability and service life.

CG-731 Funding and Acquisition Processes

As the boat program sponsor for the USCG, CG-731 controls the OE funds that may be applied to boats. This includes funds both for boat maintenance and for the procurement of boats under the allowed threshold. For a relatively simple or small number of assets, CG-731 will handle the procurement from their office. If these procurements rise to the level of a major or nonmajor system, CG-731 follows the same documented procedures as noted for CG-9325. For systems that fall under the nonmajor threshold, CG-731 follows a unique process. A variety of

---

[6] Commandant Instruction Manual 16114.4B, 2012.

documents are reviewed to support the acquisition, with ultimate sign-off authority occurring with the CG-731 PMO. The documents reviewed include

- requirements documents
- acquisition planning and forecasting
- market research
- specifications
- statements of work
- independent government cost estimates
- contracting officer's representative duties/responsibilities
- proposal evaluation plans
- procurement requests.

Analysis of Current Boats Acquisition

This review of the USCG's current boats acquisition system, along with interviews with USCG SME, suggests three areas of primary concern for a future CG-9325 boats program. These areas are funding, program structure, and project staffing. Each is discussed in more detail below.

Funding

Boats acquisitions funding currently is split between AC&I and OE. The AC&I appropriation is subject to greater scrutiny and internal and external oversight, but it has little flexibility across similar projects or product lines. In an effort to control budget growth and impose greater accountability over all government acquisition spending, the Office of Management and Budget and congressional appropriations staffs require greater justification at multiple levels and are very reluctant to increase the top line for AC&I funding. This has a natural "trickledown effect," creating greater competition between programs vying for the same appropriation. Those programs that have higher visibility or garner greater political support tend to be prioritized. An unintended consequence of this process is program lag—the extra time that it takes to move through the acquisition process to final delivery to the field user. This is somewhat ameliorated by a five-year allowance for AC&I monies.

Conversely, the OE appropriation is strictly applied to each fiscal year and enjoys much less line-item oversight and much greater flexibility across programs. However, the OE appropriation can vary internally from year to year and is subject to internal priority justification. The USCG budget model drives OE apportionment, so more-stable recurring expenses are more secure than one-time expenses. This affects small boats acquisition, as it becomes exceedingly difficult to plan only three to five years out into the future. Due to standardization efforts, the population of some types of boats requires stable multiyear funding to keep pace with asset obsolescence and to realize efficient, fully staffed programs.

We continue our discussion of benefits and challenges associated with different funding strategies in our examination of other acquisition organizations in Chapter 3. Assessments of possible PMO structure and funding strategy combinations are supplied in Chapter 4.

Program Structure

Interviews with USCG SMEs indicate two concerns regarding program structure: establishing points of contact and acquisition rigor. Stakeholders outside the acquisition community expressed concern that the current organizational structure can make it difficult to know who to contact if they have issues with a specific asset, as it is not always clear which staff are assigned to a specific boats acquisition. This is made more difficult by the split between CG-9325 and CG-731; in some cases it is unclear which office is directing procurement.

Another concern stemming from a divided program structure is acquisition rigor. USCG SMEs perceive that acquisitions funded through AC&I dollars are more rigorous than those funded through OE funds. However, follow-up interviews and examination of documents suggest that acquisition rigor is driven primarily by acquisition category rather than funding source. Both offices, CG-731 and CG-9325, must follow the same USCG procedures when acquiring major and nonmajor acquisitions. These processes do not vary by office, and they head to the same higher-level signatories, regardless of office of origin.

There is more ambiguity in how programs that fall below the nonmajor threshold are managed. Interviews with several SMEs noted that guidance for these projects varies from project to project, even within an office. The driving force in these acquisitions appears to be completing the required documentation for a contracting officer to approve the establishment of a contract. As documentation changes over time, and as these reviews tend to be internal to an office, documents and artifacts produced vary from acquisition to acquisition. Further guidance and standardization in this area would provide clarity and rigor to nonmajor acquisitions.

Project Staffing

A final concern is the ability to staff projects within boats acquisition. CG-9325 has seen its number of acquisition projects and total number of assets in its portfolio increase over time—without an accompanying change in staffing. Along with our interviews with USCG SMEs outside the boats acquisition office, this suggests that staffing has not kept pace with workload demands. Given its matrix-based structure, CG-9325 need not add a completely new staff for each new project. However, given the number of assets in the USCG's boat fleet, concerns over staffing levels persist. These concerns are exacerbated by split funding; although many boats are funded via OE, staffing billets within the acquisition directorate are accounted for based on a program's AC&I funds. This creates management complexities because of regulations about the amount of time personnel in AC&I billets can use to support OE projects (and vice versa). It might also pose a challenge to boats acquisition if new boat types require future acquisitions without additional billets. Additional staffing-related concerns include staff experience with acquisition and resident subject-matter knowledge of boat missions and operations. However, the overall staffing level was most often brought up in interviews and discussions with CG-9325.

3. Acquisition Work Requirements, Procedures, and Processes in Other Organizations

This chapter describes and compares seven organizations—three internal to the USCG, four external—that fulfill acquisition functions similar to those the USCG needs for its boats acquisition program. We sought to understand how these organizations are structured, funded, and staffed to see which of their management aspects could inform CG-9325's structure and processes. We paid close attention to management methods that underpinned an enduring program of varied projects managed within the same office. External to the USCG, we examined a limited subset of programs that manage boat forces within DHS and other parts of the U.S. government. This subset is representative of broader organizational acquisition efforts.

Our comparisons between CG-9325 and other organizations suggest that CG-9325 is different in a number of important ways. Nonetheless, the USCG can learn from these other acquisitions organizations for its own boats acquisition.

Analogous USCG Organizations

Within the USCG, we focused on three organizations that face similar, but certainly not identical, acquisition challenges to those CG-9325 confronts. These are CG-9323 (ISVS); CG-9335, the C4ISR program; and the EHRA program, part of CG-9332. All of these are enduring programs in the acquisitions directorate that manage a variety of projects simultaneously. The ISVS program within the surface acquisition directorate manages the Service Life Extension Program (SLEP) refits using the USCG Yard or private commercial shipyards for multiple larger vessels simultaneously. CG-9335 C4ISR is a program within the command, control, communications, and computers acquisitions directorate that provides C4ISR equipment to multiple classes of new USCG cutters. EHRA is a part of the work ongoing in CG-9332 that is considering how to manage the USCG's medical records. These organizations procure many fewer assets than CG-9325, but they face the similar challenge of managing multiple disparate projects within a single PMO and manage projects that could be considered enduring in nature. As such, we consider these organizations to be the PMOs that are most analogous to CG-9325 within the USCG and possible sources of best practices for a future CG-9325.

Funding

ISVS and C4ISR projects are funded through AC&I funds, but via different mechanisms. ISVS receives a single direct AC&I appropriation that it has the authority to distribute among the active projects it manages. C4ISR is also funded with AC&I funds, but it receives them from the cutter projects they support rather than directly as a single apportionment from the USCG's general AC&I appropriation from CG-82. EHRA is part of a broader PMO funded via AC&I

funds, but its efforts are currently funded through OE funds. The program is still considering acquisitions options, some of which can be funded through AC&I funds and some which can be funded through OE funds. In general, there is concurrence within USCG programs that AC&I funding helps stabilize acquisition processes, as is to be expected with multiyear funds, but this process is not without its challenges. AC&I funds must be allocated for each individual project—and this allocation means a multiyear process of requesting, justifying, securing, and defending funding. This challenge could be addressed by combining AC&I funds for several projects into a single appropriation, as is currently done with ISVS.

Funding acquisitions across multiple related projects in a single AC&I appropriation provides several benefits to a program. It provides stability of funding over time, while maintaining flexibility to move funds to projects as required by current USCG needs. This can be especially important for projects, such as boats, that require planning for future acquisitions when some of the assets have not even reached the halfway point of their useful lives. Another aspect of flexibility offered by a direct appropriation is stability in paying for program management, administrative, and other support staff. Staff could be capable of working on any given project within the office, not restricted to the single project that funds their contract. This arrangement appears to be unique to ISVS among USCG programs, and whether it can be replicated in other PMOs is unclear.

Organizational Structure

We consider three types of internal PMO structures that may be applicable to the USCG's needs: *project-based, functional-based*, and *matrix-based*. These are standard program structures commonly used in program management and are well defined in textbooks and related expert literature.[1]

Project-based PMOs have self-contained internal divisions tied to specific projects. Each internal team has all of the skills and expertise—including project management, engineering, and logistics expertise—to manage a specific project from start to finish. For example, a project-aligned CG-9325 would have a project team devoted solely to the Response Boat–Small project, another to the Over-the-Horizon project, another to the Long-Range Interceptor project, and so on for all the projects managed by CG-9325.

Functional-based PMOs invert this arrangement. Rather than having a specific team for each project that manages that project from start to end, functional-based PMOs have teams organized around a specific task or substantive area, and a project is passed from team to team across the acquisition process. For example, one team would focus on system engineering and, once a

---

[1] See Kerzner, 2003, for in-depth descriptions of traditional (functional), product-centered (project-based), and matrix programs; Jay R. Galbraith, "Matrix Organization Designs: How to Combine Functional and Project Forms," *Business Horizons*, Vol. 14, No. 1, February 1971, for a review of the development of these organizational forms; Robert C. Ford and W. Alan Randolph, "Cross-Functional Structures: A Review and Integration of Matrix Organization and Project Management," *Journal of Management*, Vol. 18, 1992; and J. Rodney Turner and Anne Keegan, "The Versatile Project-Based Organization: Governance and Operational Control," *European Management Journal*, Vol. 17, No. 3, 1999, for a synthesis of advantages and disadvantages of these organizational types.

design is selected, would pass contracting and other production duties to different separate team (and so on).

Matrix-based PMOs combine some aspects of project and functional PMOs. In this organizational structure, each project, or group of related projects, is assigned an APM to manage it across it across the acquisition process. Additional personnel within the PMO are organized into groups similar to those found in a functionally aligned PMO; however, each individual is assigned to a particular program and reports to the APM. The APM might have a few other staff members assigned permanently to the project, but, for the most, he or she would draw on staff from separate functionally organized groups. For example, the APM responsible for the Long-Range Interceptor would draw on the same pool of technical and logistics expertise as the APM for the Response Boat–Small project.

The ISVS, C4ISR, and EHRA PMOs are matrix-based organizations, but each is organized quite differently. ISVS has a small core staff, with some personnel matrixed into the office. This staff is largely organized around a matrix structure, with individual APMs for each active acquisition project drawing on other functionally grouped staff as needed. The C4ISR PMO is organized differently, with most of its personnel working across the organizations responsible for cutter acquisitions. These billets are placed with what could be considered the customer organizations, but they ultimately report to the C4ISR PMO. EHRA draws personnel from the broader CG-9332 and is still in the process of adding staff specific to its project.

The different product each organization produces likely drives this variability. ISVS deals with vessels that have already been procured and can manage all aspects of the SLEP process. On the other hand, C4ISR provides systems that must be integrated as part of a broader acquisition; it is logical to establish an organizational structure that equally integrates the PMO as well. Similarly, EHRA is at an early stage in the acquisition process, having not yet selected an acquisition option, and as such has a relatively new organization.

Staffing

Staffing levels vary significantly in ISVS, C4ISR, and EHRA. ISVS has a headquarters staff of seven: five core personnel with two "matrixed" staff. There are also 25 people at the USCG Yard who act as a project residence office and address quality control and delivery issues. The C4ISR PMO has 74 staff, split roughly equally between civilian, contractor, and military personnel. They are spread across multiple sites and, as previously noted, located within different headquarters offices. EHRA has a staff of five, with plans to build to a staff of 12 by the end of fiscal year 2017; 80 percent were to be military staff. Because all of these programs are funding these billets with AC&I funds, they face the usual challenges of USCG acquisitions, including justifying AC&I billets for organizations that do not directly receive AC&I funds and difficulties in matching workload to funding, given long lead times in obtaining project specific funds. CG-9332 shares a challenge with CG-9325: managing OE-funded projects with only AC&I billets. In these cases, personnel are expected to work at least 51 percent of their time on AC&I projects, creating further time-management challenges.

Summary of Findings: Internal Organizations

All the internal USCG organizations we examined have similarities with and differences from CG-9325. Two of these programs are funded entirely by AC&I funds, without the OE fund dichotomy that CG-9325 faces. C4ISR receives its funding from cutter procurement, just as CG-9325 receives some of its AC&I funding for cutter boats via similar methods, while ISVS receives a single direct AC&I appropriation. CG-9325 is organized more like ISVS or EHRA than C4ISR, with project-based APMs and functional staff, rather than staff matrixed into other organizations. Finally, both ISVS and C4ISR also have more staff than the current boats program. These differences suggest that increases in staff would be appropriate to align an enduring boats program with current enduring acquisition programs within the USCG. For ISVS, a large portion of these staff reside in a project residence office that addresses issues with production directly, freeing headquarters staff for other acquisition tasks.

Figure 3.1 presents these four organizations by their funding source, organizational structure, staffing size, and the types of staff involved in a program. The figure axes align with the previous discussions of whether a funding type is present in a program and PMO structure. The size of the pie charts for each office is proportional to the total number of staff in a PMO. Finally, the wedges in each pie represent the percentage mix of civilian, military, and contractor personnel in a PMO. CG-9325 is similar in structure to other USCG organizations and is the only mature acquisition program that draws both AC&I and OE funds.

Figure 3.1. Summary of Analogous USCG Organizations

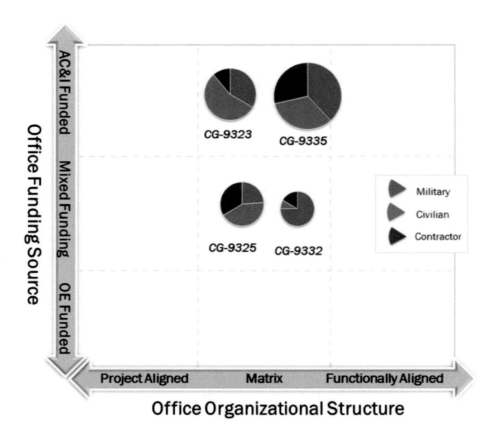

Analogous Organizations External to the USCG

We examined four organizations that acquire boats for other government agencies: the Army Corps of Engineers Marine Design Center, CBP Air and Marine, NOAA Offices of Law Enforcement and Fisheries, and NAVSEA PMS 325. The Army Corps of Engineers manages 1,000 significant floating assets and 3,000 additional assets, with an average service life of 30 to 40 years each. CBP Air and Marine manage approximately 200 boats of six different classes. NOAA Law Enforcement and Fisheries combined manage a fleet of approximately 300 boats. NAVSEA PMS-325 manages a total inventory of 3,000 boats of several classes, procuring 60 to 70 boats a year with an average service life of 12 years each. All of these offices procure boats with varying degrees of similarity to those acquired for the USCG, but they represent the most similar sets of asset classes across the federal government. These organizations could potentially inform CG-9325's structure and funding strategy.

Funding

Like CG-9325, most other organizations manage multiple funding streams to purchase boats. The Army Corps of Engineers funds projects in two ways: directly through corps funds or directly from sponsors. There is no ambiguity in which funding stream a particular process draws on, and both are managed out of the same PMO. NAVSEA PMS-325 draws approximately $45 million a year in acquisition funding from several sources. These sources differ by the complexity of the project; however, as PMS-325 is solely responsible for all boats acquisition for the Navy, this does not seem to cause confusion or uncertainty in the acquisition process. CBP Air and Marine funds all of its projects through AC&I. Each CBP Air and Marine boats acquisition project must compete against all other CBP Air and Marine acquisitions, but recently such projects have not fared well given the advanced ages of existing platforms and the need to upgrade to meet new requirements. NOAA's Law Enforcement and Fisheries allocate boats acquisition funds to regions; the regions then decide how to prioritize boat purchases. Many of NOAA's small boats acquisitions are highly specialized and unique to individual science grants.

Organizational Structure

External boat acquiring organizations tend toward a matrix structure. The Army Corps of Engineers is organized into a design branch and a project management branch. For each project, a team of one to five people, composed of personnel from both branches, is built to manage the project. This is another version of a matrix-based organization, with particular projects drawing on technical and other expertise from a stable pool of staff. CBP Air and Marine uses a matrix-based organizational structure. A boats acquisition project will have a PM assigned to it permanently, and that PM draws on other members of the acquisition directorate to fulfill key program tasks. The NOAA offices, as mentioned earlier, manage boats acquisition at the regional level. NAVSEA PMS-325 staffs projects with a core team of five staff for procurement, including a lifecycle manager, in what we might consider a slightly project-based matrix organization.

Most of the external programs we interviewed are similar to CG-9325 in that they do not manage the programs from "cradle to grave" (i.e., throughout the entire acquisition lifecycle). They typically have some level of involvement at all stages but are not the primary manager of the boat assets after acquisition. Ownership is generally transferred to the user. Given this model, one might expect to see more function-based organizational structures, but the organizations interviewed tended toward matrix-based structures. Many of the interviewees emphasized the criticality of effective communication between the acquisition office and the user office to achieve program success. The matrix structure offers flexibility, but it still provides stability and key interfaces for acquisition workforce and end users to communicate effectively throughout the lifecycle.

Staffing

Staffing sizes of boats acquisition organizations vary among external organizations, as does a program's ability to draw on technical expertise outside its immediate staff. The Army Corps of Engineers Marine Design Center is staffed by from 28 to 33 personnel; staffing levels have been within this range for the past 20 years. A typical PMO in CBP Air and Marine will have one full-time PM and six other part-time staff support, split between contractor and government personnel. CBP Air and Marine has also established the ability to reach out to the Navy for engineering and technical requirement support. NAVSEA PMS-325 has a core staff of approximately five, not including the one to two dozen person-years of technical and engineering staff drawn from broader NAVSEA engineering support services. The NOAA offices do not handle acquisition through a central office, which makes it difficult to assign staff to any particular acquisition. In general, one or two personnel oversee regional efforts, and each boat type has a designated boat coordinator that manages individual vessels.

Summary of Findings: External Organizations

No external organizations procuring boats face exactly the same missions and acquisition challenges as the USCG, but some of these organizations' process can be used to improve existing USCG processes. Many organizations manage multiple types of funds, as CG-9325 does. For Army and Navy organizations, fund management is made easier by being the sole organization able to procure boats within their service. Notably, the other DHS organization procuring boats (CBP Air and Marine) does not have multiple types of funding, relying on AC&I for acquisitions. Matrix structures are dominant among boat procurement organizations in external organizations that procure boats through a central office. Similarly, staffing among organizations that manage acquisition for fleets of over 1,000 assets appears to be relatively stable between 20 and 30 personnel.

Figure 3.2 compares these external boats acquisition organizations with CG-9325 in terms of the degree of centralization involved in boats acquisition, PMO structure, staffing size, and staffing mix. In the chart, the vertical axis represents the degree of centralization, which varies considerably among the organizations we examined. We do not use funding type as an axis to distinguish between AC&I and OE funds, as this comparison is not relevant outside of the USCG and other DHS organizations. The remainder of the figure follows the concept presented in Figure 3.1. Note that CG-9325 is closer in structure and fleet size to the Army and Navy organizations than it is to CBP or NOAA.

Figure 3.2. Summary of Analogous Organizations External to USCG

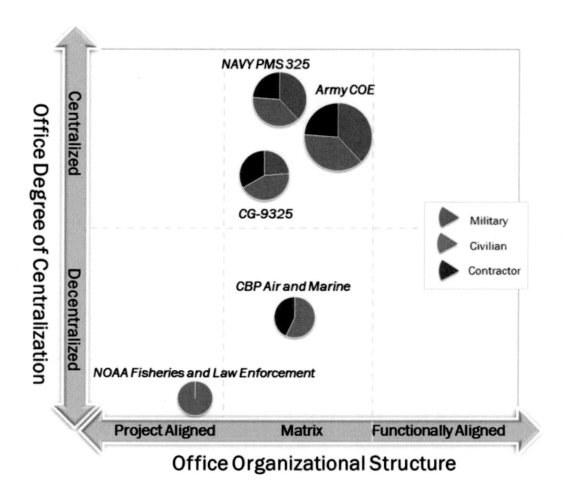

Comparing Current CG-9325 with Analogous Organizations

Comparing USCG-affiliated and external organizations that are similar to CG-9325 produces the following observations:

- **Matrix-based organizations are dominant.** We find that the matrix management structure is common among both USCG-affiliated and external programs facing acquisition challenges similar to those CG-9325 confronts.
- **Multiple funding streams are managed better under one office.** CG-9325 is the only organization examined under DHS oversight that does not meet most of its acquisitions (in terms of number of boats acquired) with AC&I funds. Some external organizations manage multiple funding streams, but they ameliorate this challenge by being the sole procurers of boats for their organization.
- **Large fleets require more staffing.** Programs that manage acquisition for fleets of more than 1,000 assets, as CG-9325 does, tend to have 20 to 30 staff (military, civilian, and contractor) at a given time.

Across the many dimensions considered, CG-9325 is different from other organizations that face similar acquisition challenges in terms of organization and funding strategy. Nonetheless, it can still draw best practices from them. Like most other organizations we examined, CG-9325 has a matrix structure for its PMO. CG-9325 is different from any other organization we examined in that it is not the sole procurer of boats within a centralized organization, and it is also the only organization that uses operating funds for boats acquisition. These comparisons, along with our discussion of the current state of boat procurement in Chapter 2, suggest that changes to CG-9325's organizational structure and funding strategy should be considered. In the next chapter, we examine the combinations of multiple organization forms and funding strategies suggested by these comparisons to assess the impact of their adoption on a future USCG boats acquisition program.

4. Comparative Assessment of Boats Acquisition Programs

This chapter describes and compares alternative program structures and funding strategies for the CG-9325 boats acquisition program. To facilitate comparisons, we assess the benefits, challenges, and risks to the USCG associated with each option. For our assessment, we conducted a qualitative analysis based on interviews with SMEs and our own research. We developed assessment criteria, developed three alternatives for both program structure and funding strategy choices, and evaluated each alternative's ability to meet each criterion. Although all combinations of program structure and funding strategy discussed here could be made to function by the USCG, some combinations add greater value, and others introduce more challenges.

Our assessments suggest that a matrix-based PMO pursuing a mixed funding strategy focused on AC&I funds would be most appropriate for CG-9325.

Assessment Criteria

To compare different options for PMO structure and funding strategies, we assessed alternatives across a range of challenges that the prospective boats acquisition program will likely need to address: funding, acquisition rigor, and project staffing (as identified in Chapter 2). We do not explicitly assess the ability of these potential structures and strategies to meet cost, schedule, and performance criteria. Although meeting these triple constraints is obviously critical, we did not find doing so to be a discriminator in assessing office structure and funding strategy alternatives. Given that USCG personnel would execute any structure or strategy, we find no reason to suspect that any option chosen would be unworkable on these dimensions. Instead, we expect that some options would present more challenges to those personnel when compared with the alternatives.

The challenges discussed here are not exhaustive; rather, they are key challenges made apparent from program history and program strategy and goals or were noted by program stakeholders during interviews. The previous chapters suggested that addressing four challenges would be important to the CG-9325 program; we use these challenges as criteria by which to evaluate each alternative strategy:

1. **Ability to add additional projects:** History indicates that when the need to acquire a new type of boat arises, this need will be added to the responsibilities of the boat's PMO. Regardless of the specific project, currently unforeseen boats acquisition will need to be managed by an enduring CG-9325. A program structure or funding strategy is effective on this metric if it minimizes the challenge of adding new acquisition projects.
2. **Flexibility:** This challenge refers to the capacity of any future PMO to innovate or otherwise change course once a project is under way or to switch between projects within the office. Successful acquisition projects require a degree of flexibility to adapt to

changing circumstances during the acquisition timeline. These circumstances can involve changes in funding and a need to prioritize one project over another, a need to address emergent requirements that shift the course of a project, or complications that arise during the acquisition or procurement process. Some organizational structures and funding strategies are better able to adapt to these and other challenges. A program structure or funding strategy is deemed effective on this metric if it can be flexible during the acquisition projects it manages.

3. **Ability to interface with other stakeholders:** Another challenge is the need for a boats program to interface with a variety of other USCG stakeholders. During the course of an acquisition, CG-9325 interfaces with a variety of other USCG offices to ensure that the system meets the needs of all parts of the organization. This ranges from coordinating across the organization as requirements are written to ensuring that contracts contain provisions required for later sustainment concerns. Executing these interfaces, whether through staff points of contact or formal processes, is vital to any acquisition program.

4. **Staffing:** The final primary challenge is the impact of a structure or funding strategy on PMO staffing. This includes the number of staff; the distribution of military, civilian, and contractor staff; and the relative composition of program management, logistics, and technological skills within the office. This dimension also addresses the stability of staffing over time and the general staffing demands indicated by program structure and funding strategy.

We score each potential program structure/funding strategy against these four challenges, assigning a score of 1, 2, or 3. A "3" indicates that the potential program structure/funding strategy is very likely to be able to address the specific challenge.

Some of these challenges could potentially be mitigated by separate efforts: For example, establishing a long-term standardization and recapitalization plan would remove much of the difficulty of adding multiple projects or the flexibility required to deal with shifts between projects. However, as such a plan is not presently in place for all types of boat assets, and as no plan will ever fully account for all of the challenges a PMO will face, we assess performance against these dimensions based on the current boats acquisition situation. We use assessment criteria that focus on challenges that CG-9325 is likely to face in some capacity regardless of the future program, based on research and discussions with other enduring programs.

Office Structure Alternatives and Assessment

We consider three alternatives for CG-9325 office structure: project-based structure, functional-based structure, and matrix-based structure. We describe each of these below, along with some of their advantages and issues. Then, we evaluate each using the four challenge criteria just described. A summary of the structures are shown in Table 4.1 below. Each structure is also described in more detail in Chapter 3.

Table 4.1. Office Structure Alternatives

Alternatives	Description
Project-based structure	Internal divisions are tied to specific projects. Each project has a team that has all of the skills and expertise to manage a project across the acquisition cycle.
Functional-based structure	Inverts the project-based arrangement and represents a traditional view of organizational design. Rather than having a specific team for each project, teams are organized around a specific task.
Matrix-based structure	Combines project and functional-based structures. A single APM is responsible for managing a project across the acquisition process, and the rest of the project team is drawn from common functional groups. A small core staff for each project alongside areas of functional expertise allows for adaptability between projects.

Project-Based Structure

A project-based structure is based on internal divisions tied to specific projects. Each project has a team that has all of the skills and expertise to manage a project across the acquisition cycle. For example, a project-aligned CG-9325 would have a project team devoted solely to the Response Boat–Small project. In this structure, staff remains in the same project over time. This provides a PM complete line authority over individual projects, allowing project management to be flexible and responsive as the project changes. Allowing a PM direct authority over his or her entire staff provides clear communication channels and enables quick decisionmaking. Changes can be driven by the customer or other stakeholders, but, regardless of the source, the project is organized to be able to respond to these changes. Furthermore, keeping staff on the same project over time allows them to accrue institutional knowledge across a project's lifecycle.

Additionally, the stability of a project team eases interfaces with other stakeholders in the USCG. Having a single point of contact for other offices can ease coordination, especially as familiarity increases and the team develops knowledge of other organizations' needs (this knowledge also can be applied to future projects).

The challenges associated with a project-based PMO center on the number of staff required. In a project-based PMO, the organization cannot leverage the technical expertise of one individual for more than one project. This duplication removes the ability of the PMO to benefit from economies of scale, and personnel may be kept on projects after they are needed. In a billet-constrained environment, there is substantial risk associated with a structure that needs additional staff to function. This staffing issue also creates a risk when additional projects are added to the office.

Based on this assessment, we would expect a project-aligned PMO to excel at addressing challenges related to flexibility and interfacing with stakeholders, as a single team is responsible for a project at all stages of the acquisition process. This structure would struggle to add additional projects to the PMO, as each new project would require a completely new project team's worth of billets. Similarly, this structure is likely to struggle to reach an adequate number of staff, and to arrange for each team to have all requisite skills, given the resource-constrained environment in which boats acquisition occurs.

Functional-Based Structure

Functional-based PMOs invert the project-based arrangement. Rather than having a specific team for each project, functional-based PMOs have teams organized around a specific task; a project is passed from team to team across the acquisition process. This arrangement provides several staffing benefits. By focusing on functional areas, staff members are able to develop deep skills in their areas and produce economies of scale in key areas of expertise. As the same group of staff executes a function for all projects the PMO manages, it is easier to implement lessons learned in future projects. Furthermore, it is also easier to standardize PMO practices, since the same group conducts similar work across projects. This stability of staff also eases the process of interfacing with other USCG stakeholders, as once again there is a stable group of contacts. Finally, it is much easier to add projects to a PMO of this structure. Theoretically, the PMO is staffed to accommodate the project, and it can simply flow from team to team as the project moves through its lifecycle.

However, functional-based PMOs can encounter challenges. This organization type requires extensive coordination across teams within the PMO to ensure that handoffs are successful, requiring additional senior leadership attention and either organic coordination or greater centralization of authority. This need for coordination also makes it more difficult for projects to adapt quickly, since each change requires further coordination between functional areas. Finally, a functional-based PMO also raises staffing concerns. In this case, it is not the number of staff that is a problem but rather ensuring that all functions are appropriately tasked at any given point in time. For example, if multiple projects come into the PMO at the same time, those working on the initial aspects of the acquisition would have more work than they could reasonably accomplish, even if the PMO itself is fully staffed. Significant challenges can emerge when synchronizing work across various teams with competing priorities and requirements. This can introduce project delays or quality issues (or both) if tasks cannot be logically sequenced due to lack of key team member availability.

Based on this assessment, we would expect a functional-based PMO to excel at adding new projects to a program, as there is no need for additional staff (assuming appropriate planning). A functional-based PMO would have both benefits and challenges in terms of stakeholder interface, as specific points of contact would be able to reach out to a functional team. However, there would be no single point of contact responsible for the entirety of a project. Fewer staff members are required in this arrangement than for a project-based PMO, but managing workloads across those staff would be more difficult for senior leadership. Finally, this structure would struggle the most with flexibility, as project functions are confined to separate project areas, and no one (save senior program leadership) has full authority and responsibility for any particular project. This stovepiping effect makes it difficult for projects and programs to change course without extensive coordination between functional areas.

Matrix-Based Structure

A matrix-based structure combines the two previous structures by using a single APM responsible for managing a project across the acquisition process and drawing the rest of the project team from common functional groups. This combination provides several benefits, as the APM has control over technical aspects of the program, but personnel are assigned to the project. This makes it possible to tailor the staff mix to meet a project's specific needs. Each person assigned to a project also has a "home base" in a functional area, so there is less concern about what happens when the project ends. This type of organization was developed to address the weaknesses of other organizational structures by combining specific aspects of the project-based and functional-based organizations. Having a small core staff for each project, alongside areas of functional expertise, creates a flexible organization that is adaptable within projects.

A matrix-based structure is not without its challenges. Lines of authority in a matrix-based organization can be difficult for staff to manage, with multiple reporting chains and a need to split time between projects and different PMs. The "two-boss" problem is very common in matrix organizations and can lead to conflicting direction. This confusion can also extend outside the organization, as outside stakeholders could lack a stable point of communication (outside the PM) for each project. Consistency across projects can be problematic in a matrix-based organization, which can also lead to issues in communicating with and providing products to external stakeholders, as they lack stable contacts within the boat program across projects.

Based on this assessment, we would expect a matrix-based PMO to be able to address challenges associated with adding additional projects and flexibility during acquisition by combining the best of project- and functionally based PMOs. New projects could be added by adding a new APM and drawing on existing staff in functional areas; this APM would have the authority to make changes as needed to the project once it is underway. This structure would offer benefits and challenging for interfacing with stakeholders, as they could contact a single APM to contact, but there would be no clear points of contact for functions beyond program management. Finally, staffing remains a concern in this structure, given the established matrix problems of dual reporting chains and the complexities of managing staff time between multiple PMs.

Office Structure Assessment

None of these organizational forms is perfect: Each excels in some areas and presents challenges in others. Table 4.2 summarizes the scores that each office structure receives when assessed against the four challenges. An office structure likely to be very able to address that challenge receives a 3, somewhat able receives a 2, and not very able receives a 1. In the case of staffing, no office structure received a 3, as this challenge will be present in all office structures we considered.

Table 4.2. Summary of Program Office Structure Assessment

	Project	Functional	Matrix
Add additional projects	1	3	3
Flexibility	3	1	3
Interface with other stakeholders	3	2	2
Staffing	1	2	2

NOTE: 3 = very able to address challenge; 2 = somewhat able to address challenge; 1 = not very able to address challenge.

From this summary, we observe three key insights. First, matrix-based organizations are better able to meet challenges than the other organizational structures considered. Second, all organizations, including matrix-based, face some challenges, particularly in the area of staffing. Finally, project-based and functional-based organizations face a similar level of challenges but have difficulty addressing different issues. If either of these options is selected, these challenges will require further attention.

Overall, taking these challenges in total indicates that a matrix-based approach is better able to address these challenges, although not markedly so. Importantly, it lacks the low values ("1") for some criteria that we found for the two approaches, and is able to at least somewhat address any of the challenges considered. Given the persistence of these challenges in boats acquisition, an approach that has a substantial limitation in any respect is likely to generate more problems in the future than one without such limitations.

Funding Strategy Alternatives and Assessment

As in the previous section, we describe and assess different strategies for a future CG-9325. The strategies we consider here pertain specifically to funding for both assets managed by an office and the office staff. A summary of the strategies are shown in Table 4.3 below.

Table 4.3. Funding Strategy Alternatives

Alternatives	Description
Fully AC&I Funded	AC&I is a multiyear process that normalizes the process of obtaining staffing billets. Fully funding the boats program through AC&I funds would normalize the process of obtaining staffing billets, which are largely based on AC&I funds within CG-9. The multiyear nature of AC&I funds also ensures staffing stability over time.
Mixed Funding: OE Primary	This method would use OE funds in allowed cases, with AC&I filling in only for above-threshold procurements (essentially the current process in USCG boats acquisition).
Mixed Funding: AC&I Primary	This strategy primarily utilizes AC&I funding to maintain staffing levels; funding for projects over the OE threshold comes directly to the boat program. OE still may be used for smaller projects and for contingent and emergent requirements.

30

Fully AC&I Funded

Fully funding the boats program through AC&I funds, as is the case in most other acquisition programs within CG-9, would normalize the process of obtaining staffing billets, which are largely based on AC&I funds within CG-9. The multiyear nature of AC&I funds also ensures staffing stability over time. The ability to access these funds across years also helps the program navigate the rigors associated with major acquisition projects.

The primary challenges associated with fully funding the program through AC&I are the time required to obtain the funds and the ability to secure long-term funding. AC&I funds are directly appropriated and must be part of a budget. It can therefore take several years from initial request to obtaining executable funds. This can prevent the timely addition of projects. This challenge can be partially mitigated with a comprehensive acquisition plan based on useful life projections, as discussed further in Chapter 5. This challenge could be mitigated if AC&I funds were appropriated for general boat procurement without regard to a specific process. However, this would require changes to existing USCG practices outside the scope of CG-9325.

We would expect a strategy of fully funding boats acquisition with AC&I funds to excel at meeting challenges related to staffing and interfacing with stakeholders, given the stability five years' worth of funds provides to a PMO. It would offer both challenges and benefits to flexibility and create challenges to adding additional projects (based on the long lead-time required to obtain new AC&I funds from the date they are first requested).

Mixed Funding: OE Primary

Fully funding a boats program with OE funds is not strictly possible, as there are threshold limits on which boats can be procured with OE funds. We refer essentially to the use of OE funds in allowed cases, with AC&I filling in only for above-threshold procurements. This is essentially the current USCG boats acquisition process.

Funding CG-9325 with OE funds presents different benefits and challenges to the USCG. As one-year funds, OE funds are more flexible than AC&I. Presently, the USCG is authorized to use up to $31 million a year of OE funds to recapitalize boats, as long as the price of the new boat is $700,000 dollars or less and can be justified as a "contingent and emergent requirement." (This exception came about because the USCG supports Department of Defense contingency operations.[1]) Through using this pool of money, it is easy for the USCG to add new procurements or to shift funds between projects in a way that cannot currently be done exclusively using AC&I funds. This flexibility enables the USCG to be responsive to emerging needs.

There are a few concerns and issues associated with an OE-centric approach to acquisition. As one-year funds, OE procurements place pressure on program staff as they attempt to complete the documents and artifacts required in a rigorous acquisition process. Furthermore, even if boat procurements can be executed with OE funds, staffing levels with in CG-9 are still based on

---

[1] See Pub. L. 113-6, 2013, p. 127.

AC&I funds. This further complicates the process for program staff, as those on AC&I billets are expected to spend a majority of their time on AC&I projects, even if most boat procurements are funded with OE funds.

We would expect a mixed funding strategy focused on OE to be able to meet the challenges of adding new projects and flexibility in acquisition, as by their nature OE funds are able to be used reactively. This same reactivity can lead to challenges interfacing with other stakeholders that often depend upon a longer process to ensure their equities are met during an acquisition. Finally, as AC&I funds drives staffing billets within CG-9, this strategy struggles to address that challenge.

Mixed Funding: AC&I Primary

A mixed funding strategy can take several forms, depending on the mix of funding types involved. Here, we consider a mixed strategy that primarily focuses on AC&I funds. This has three criteria. First, the boats program receives sufficient AC&I funds to maintain staffing levels. Second, AC&I funds for projects over the FRMM OE threshold come directly to the boats program. Third, OE funds continue to be used for at least some boat procurements falling under the FRMM threshold. These assumptions are consistent with existing guidance in the FRMM, but they also acknowledge the value of the exception made for the USCG to purchase boats using OE funds.

A funding strategy based on a mix of AC&I and OE funds has some clear advantages over a strategy focused on only one type of fund in that it leverages the best of both funding streams—the stability of AC&I funding and the reactive ability that comes with OE funding. The desirability of this strategy could vary based on the mix of funds involved, but as long as enough AC&I funds flow into the boats program to support staffing concerns and give stability to multiyear acquisitions, this arrangement faces fewer concerns than the other funding strategies considered. Managing the issue of AC&I billets executing OE procurements is one challenge that would remain as long as any OE funds are used to execute acquisitions. Another challenge that will be present as long as boats are procured via two different funding mechanisms is the need to manage accounting and oversight of two different funding mechanisms.

Funding Strategy Assessment

As in the case of office structure alternatives, none of the funding strategies considered is perfect. Each excels in one or more areas but presents challenges in others. Table 4.4 summarizes the challenges for each strategy, following the same process used for Table 4.2.

Table 4.4. Summary of Program Office Funding Assessment

	AC&I	Mixed-OE	Mixed-AC&I
Add additional projects	1	3	3
Flexibility	2	3	3
Interface with other stakeholders	3	1	3
Staffing	3	1	2

NOTE: 3 = very able to address challenge; 2 = somewhat able to address challenge; 1 = not very able to address challenge.

Overall, the analysis suggests that a mixed funding strategy with a focus on AC&I funds presents the fewest challenges of the strategies considered. This mixed funding strategy best meshes the flexibility of the current funding strategy of prioritizing OE funds for all boats under the threshold, but provides more stability to a PMO to navigate acquisition rigor and facilitate points of contact with USCG stakeholders. Using OE funds in some capacity is key to the USCG's ability to be reactive to boat needs as they arise in the fleet, but it does little to aid program stability. This challenge can be mitigated by adding AC&I funds directly to the boats acquisition program for those boats above the OE threshold. This is contingent on the assumption that the boats program directly receives sufficient AC&I funds to maintain staffing levels. If not, other challenges would likely arise.

Summary of All Office Structure and Funding Alternative Assessments

Taking these assessments together produces nine possible office structure and funding strategy combinations. We assess the dimensions to be independent, and as such we focus on the conclusions drawn from each assessment separately. **These assessments suggest that a matrix-based PMO pursuing a mixed funding strategy focused on AC&I funds would be most appropriate for CG-9325.** Currently, CG-9325 is considered a matrix-based PMO with a mixed funding strategy that prioritizes OE funds. As such, the major change we suggest concerns funding strategy. We would expect this change to a mixed funding strategy with primarily AC&I funds to improve the ability of CG-9325 to address staffing concerns and to coordinate with other USCG stakeholders while maintaining the flexibility inherent in the current funding strategy.

Of course, changes could be made to the matrix-based organization as well. For example, while the current CG-9325 has a single APM for each project, a future CG-9325 might instead assign an APM for broad classes of boats (e.g., response boats, cutter boats, ATON boats). This APM could also be joined by a permanent technical lead and possibly other staff, creating a stable group of core project staff with broad expertise in a type of boats. This assignment could ease coordination with other stakeholders, as the points of contact would be consistent across entire broad groups of boats, rather than varying for each boat acquisition. Similarly, a mixed funding strategy would present fewer challenges than the current strategy of prioritizing OE funds. If this strategy cannot be executed—that is, if a stable appropriation of AC&I funds cannot be obtained alongside OE funds—an entirely AC&I-based funding strategy would also present fewer challenges than the current strategy.

5. Findings and Recommendations

The USCG must manage a multibillion dollar acquisition portfolio to maintain its fleet of more than 1,600 active boats. To keep up with this demand in a cost-effective way, the USCG has determined that it needs an enduring PMO focused specifically on managing boats acquisition efforts. To help the USCG understand its options, we reviewed current USCG boats acquisition practices and assessed internal and external organizations with similar goals to identify best practices and make recommendations for the structure, funding strategy, and processes of this future PMO.

This chapter summarizes our primary findings, best practices, and recommendations for USCG leadership. We first focus on the primary areas of emphasis of the study: boats acquisition funding, program structure, staffing, and workforce composition. Additionally, we include a number of additional insights that contribute to the study findings and recommendations, including policy, strategic, and tactical recommendations. These findings focus broadly on guidance for acquisitions below the nonmajor threshold, the standardization of boats, program strategy and metrics, and collaboration in boats acquisition outside the USCG.

Boats Acquisition Funding

We recommend that the USCG consider a mixed funding strategy, using both OE and AC&I with an emphasis on an AC&I direct appropriation to the boats acquisition PMO. A review of current USCG boats acquisition management reveals that, compared with other boats acquisition organizations, the current USCG PMO, CG-9325, experiences unique challenges in executing its mission. Most notable among these challenges are ambiguities in financial guidance and a divided acquisition organization for USCG boats. A review of other acquisition organizations internal and external to the USCG suggests that the future boats program would benefit from the flexibility provided by using both OE and AC&I funds to acquire boats, with the primary focus being on AC&I funding. We further conclude that CG-9325 should receive a direct appropriation of AC&I funds to ensure stability across multiyear acquisition projects and support program management and contract administration costs. Executing the details of multiyear acquisitions, especially in the case of assets with as short a service life as many boats have, without this stability introduces unnecessary risk into the projects. That said, having the option of using OE funds when appropriate provides flexibility for procurement and enhances the USCG's ability to adapt to the changing needs of its boat fleet. Any system must balance these two demands: increased acquisition rigor and planning versus an adaptive ability. The complexities in the present system that arise from using two funding sources, including when to use which funding source and which office will execute a particular acquisition, could be addressed through further guidance from senior leadership.

Boats Acquisition Ambiguity

We recommend that the USCG consider resolving ambiguity introduced by Public Law 113-6 allowing for OE expenditures for "contingent and emergent requirements" with current practices. The FRMM guidance is clear overall, stating that OE is to be used for "unit-level recurring maintenance and repair" and "minor repair and corrective maintenance."[1] Public Law 113-6 introduces flexibility in the use of OE funding, but it also injects ambiguity and potential program risk into acquisitions programs. Using OE funding for multiyear acquisition purposes might introduce unintended program risk. The current legislative guidance provides the service an exemption to use operating funds for exigent and contingent small boats acquisition requirements. However, current practice uses these funds for any boat that falls below the established threshold. If this practice continues to represent the USCG boats funding strategy, the service may seek to revise the current language to provide the service an enduring option to use operating funds to procure small boats, regardless of exigency or contingency. Determining the legislative intent or clarifying the exemption further (or both) might help the USCG engage the congressional appropriations staff. The exemption to use operating funds provides the service needed flexibility to perform lifecycle management for its large inventory of small boats. The importance of boats to the USCG cannot be overstated. The service's boat fleet produces the greatest return on investment with respect to mission performance and operating hours. The majority of USCG missions cannot be executed without a boat. Moreover, as the current operating appropriations language is written and implemented, it introduces program risk if stakeholders and oversight staffs were to challenge the service's interpretation of the language in practice. Greater clarification from senior leadership on the ambiguity introduced by Public Law 113-6 would benefit future boats acquisition.

We recommend that the USCG consider designating one office as the sole provider of boats for the USCG to provide increased efficiencies and reduce uncertainty among stakeholders. Interviews revealed examples both inside and outside the USCG where organizations are able to function effectively using two types of funding processes by being the sole acquisition authority within their organization. These organizations manage the complexity of different funding streams by simplifying who is responsible for boats acquisition; regardless of funds, all acquisition is executed from a central point. To increase efficiency, decrease uncertainty among stakeholders, and facilitate the use of multiple funding streams, we recommend that the USCG senior leadership consider following suit and designate one office as the sole procurer of boats for the USCG.

---

[1] Commandant Instruction Manual 7100-3E, 2013.

Boats Acquisition Structure

We recommend that the USCG consider an augmented matrix structure based on boat type (instead of individual boat projects). Our review of boats acquisition organizations both inside and outside the USCG suggests that matrix-based organizations face fewer challenges than other organizational options. A matrix-based structure might be most appropriate for CG-9325. To align with the desire to standardize the boat force and manage the entire fleet more effectively, we recommend a matrix-based structure that is based on boat type, rather than individual projects. For example, whereas the current CG-9325 has a single APM for each project, a future CG-9325 might instead assign an APM for broad classes of boats (e.g., response boats, cutter boats, ATON boats), as discussed in Chapter 4. This construct would also allow for a clear point person within the Boats Acquisition office for projects that are not actively undergoing acquisition, and would better prepare for future acquisitions, replacements, SLEPs, etc. For active projects, integrated project teams can be established and include personnel from all necessary functional areas within the organization. A notional matrix structure is depicted in Figure 5.1.

Figure 5.1. Notional Matrix Organizational Chart

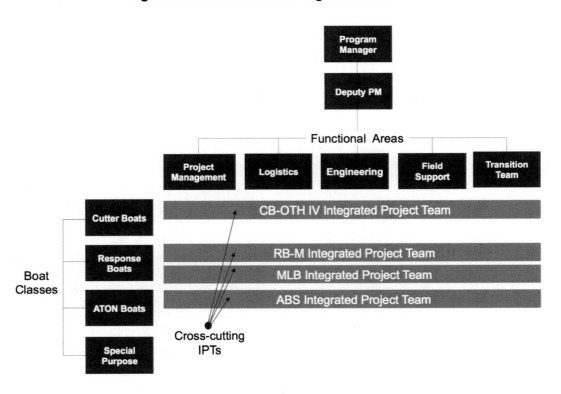

NOTE: ABS = ATON Boat–Small; CB-OTH = Cutter Boat–Over the Horizon; IPT = integrated project teams; RB-M = Response Boat–Medium.

Boats Acquisition Staffing

We observe that other comparable organizations inside and outside the USCG maintain a staff of 20 to 30 personnel and recommend the USCG undertake a rigorous analysis of anticipated workloads for the boats program and staff required to meet this workload. This observation is based on comparisons with other U.S. government organizations that manage similarly sized boats acquisition fleets. No organization we examined that manages the number and array of boat assets managed by the USCG does so with less than this number of personnel focused on acquisition. This does not include personnel that might focus solely on sustainment issues but rather is the number of personnel required to successfully complete boats acquisition. Furthermore, CG-9325 has seen the number of boat types and assets it manages increase in recent years without a corresponding increase in staff size. Related to these staffing numbers, comparisons with internal USCG organization and interviews with program staff suggest that some of this staffing should be dedicated to managing field support and transitions, tasks currently undertaken by core acquisition staff in CG-9325. Doing so would promote more-effective use of resources and build expertise in these potentially challenging areas among program personnel.

We recognize the dependency between staff size and some of our other recommendations, most notably the standardization of boats and future acquisition analysis. If a long-term acquisition strategy is developed based on useful life projections, a more-granular staffing requirement could be developed. To justify the precise staffing needs of CG-9325, a rigorous analysis of workload demands is required. Without such data, a possible distribution of staffing based on the organizational structure in Figure 5.1 is as follows:

- PM and deputy PM
- three or four personnel focused on project management, based on the types of boats the PMO is acquiring or planning to acquire
- four to eight personnel providing engineering, technical, logistics, and sustainment support as required by the complexity and volume of current and planned acquisition projects
- ten to 15 personnel to support field and transition activities
- additional personnel to support PMO administration, business support, and contracting.

If the USCG prefers to maintain the current model of two procurement organizations for boats, rather than consolidating them into a single office, one effective option is embedding personnel from CG-731 in CG-9325 (or, conversely, embedding personnel from CG-9325 in CG-731). The proposed matrix construct is especially effective for bringing in personnel from other organizations to improve the flow of communication between organizations and capitalize on the expertise individuals bring from different groups. These activities might help sufficiently prepare CG-9325 to support future acquisition efforts.

Boat Acquisition Workforce

We recommend that the USCG consider maintaining a balanced workforce in CG-9325 mixing civilian, contractor, and military personnel. A key challenge in designing an acquisition program for a military service is balancing the workforce in terms of civilian personnel (including contractors) and uniformed members. Civilians can often provide continuity and extensive experience in particular aspects of acquisition, both of which can improve acquisition outcomes. On the other hand, uniformed USCG personnel—often coming from operational roles—need to be part of the acquisition process to infuse it with a deeper understanding of current operational needs. Their frequent rotation into such positions can help raise questions that might not otherwise be considered by longstanding acquisition personnel. To the extent that some civilians are themselves former military operators, they can provide aspects of both types of benefits, though currency and fresh perspectives of rotating military personnel are still essential.

In interviews with staff members, we did not find indications of limited civilian opportunities. CG-9325 is a small PMO, and with a civilian PM and a matrix-based structure that allows for PMs for each active program, there appear to be sufficient civilian leadership opportunities relative to the size of the program. What we did find in several interviews is an emphasis on the importance of civilian positions within PMOs. One of the issues in managing a fleet with more than 1,600 assets is the need for resident expertise. Several interviewees, both inside and outside the USCG, emphasized the need for stability and continuity in key positions, such as PMs. Another point raised during interviews is that the CG-9325 leadership structure does not strictly follow the conventional military PMO hierarchical structure of leadership. Specifically, the boats acquisition program leadership structure is lower in rank compared with other acquisition PMOs. This might precipitate an unintended interoffice staffing parity issue or create the perception that the boats acquisition office is not equal in importance to the other acquisition offices.

Acquisitions Below Nonmajor Threshold

We recommend that the USCG consider establishing guidance for the prescribed acquisition process that should be followed for purchases of boats below the nonmajor ($10 million) threshold. Interviews with SMEs and literature reviews provide no specific guidance for the prescribed acquisition process that should be followed for purchases of boats below the nonmajor threshold. Recent service efforts to standardize boats, consolidate boat types, introduce configuration management discipline, and create new organizational structures that centralize support management functions have begun to change the culture of boats acquisition. For many years, the acquisition strategy was one of using a decentralized procurement and support management structure. In practice today, the CG-731 process intent for OE purchases closely mirrors the same rigorous process employed for higher-value acquisitions. However, we were unable to find written authoritative guidance that ensures consistency across multiple asset acquisitions, and we were also unable to find a formal process to promote repeatability. Establishing guidance consistent across organizations would be beneficial. At a minimum, widespread use and concerted efforts to populate a best practices database could help continuous improvement efforts and aid in the refinement of acquisition processes. Exploiting quality data acquired from a database of best practices and lessons learned could greatly assist any future efforts directed at process improvement.

Cutter Acquisition Alignment

We recommend that the USCG consider development of a Boat Capital Asset Management Plan. During our research, we interviewed leadership in CG-751 as stakeholders in the small boat program. CG-731 sees Cutter Forces as one of their primary customers, in that a large portion of the boat inventory resides aboard a fleet of approximately 250 cutters. The CG-751 organizational structure and responsibilities are analogous to CG-731, with a few exceptions. All cutters are acquired with AC&I funds, and their average lifecycle and service life are much longer. Moreover, CG-751 is also in competition with CG-731 for a portion of the AC&I appropriation. Cutter acquisitions are governed by the identical DHS and USCG policy guidance as boats, with one noteworthy exception: cutter asset lifecycle management.

Cutter asset lifecycle management is guided by the *Cutter Capital Asset Management Plan*, which is an authoritative policy document that provides guidance for the overall process of both managing an asset through its lifecycle and how a matrix-based organization facilitates a standard process to manage cutter acquisition and sustainment.[2] This approach acknowledges the complexity of asset lifecycle management and the need for multidisciplinary program support. The instruction details scope, internal stakeholder roles and responsibilities, inputs and outputs, and clear definitions or terms of reference. The discussion of scope was particularly helpful and comprehensive, providing specific program guidance in a concise and clear manner. It may be worthwhile for CG-731, in consultation with CG-9325 and other stakeholders, to develop a similar guidance document for management of the service's boat fleet.

Standardization of Boats

We recommend that the USCG consider standardization of the boat fleet to reduce risks and costs associated with acquisition and maintenance. For every boat that the USCG acquires, both operators and maintainers need to be trained to perform their respective roles. Spare parts for specific boats need to be purchased in advance, then delivered and stored where they might be needed. Every type of boat needs to have communications systems and other technologies integrated into it, and this integration can vary across different boat types. The more types of boats that the USCG has, and the more differentiated those boats are from one another,

---

[2] Mark E. Butt, *Cutter Capital Asset Management Plan (CCAMP)*, Commandant Instruction 4700.1, Washington, D.C.: United States Coast Guard, October 9, 2012.

the greater the integration costs. Perhaps a more serious matter is that operators and maintainers who are switching among multiple boat types might be prone to errors, which can risk mission effectiveness.

In this context, we find that the current range of USCG boat types is larger than may be desired. A USCG database indicates that there are 49 to 72 models of boats in the USCG's current inventory.[3] As shown in Figure 5.2, no one boat type dominates the inventory. Rather, there are 26 boat types with more than ten boats each, and even the top ten boat types collectively reflect just 67 percent of the total inventory.

---

[3] We have seen other sources suggesting slightly different numbers of boats in the inventory. This reflects, at least partly, different approaches to counting boats that are being acquired or disposed of, as well as data-latency issues regarding such boats. If we had made the extreme assumption that each boat without a designated model was of a different model from all the others by that same manufacturer, we would have many hundreds of boat models; this clearly strains intuition and conflicts with the descriptions of diverse SMEs. Ambiguity about the number of models stems from the fact that some boats are designated only by manufacturer and not by model. The high estimate corresponds to the assumption that all boats without a specified model are of the same model; the low estimate assumes that, when the model is not listed, that boat belongs to one of the models that is listed.

Figure 5.2. Types of Boats in the USCG Inventory

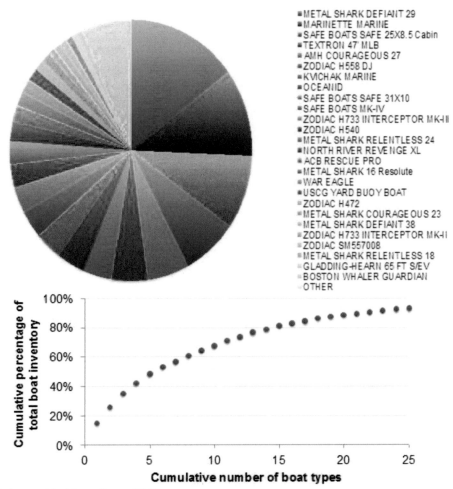

Legend:
- METAL SHARK DEFIANT 29
- MARINETTE MARINE
- SAFE BOATS SAFE 25X8.5 Cabin
- TEXTRON 47' MLB
- AMH COURAGEOUS 27
- ZODIAC H558 DJ
- KVICHAK MARINE
- OCEANID
- SAFE BOATS SAFE 31X10
- SAFE BOATS MK-IV
- ZODIAC H733 INTERCEPTOR MK-II
- ZODIAC H540
- METAL SHARK RELENTLESS 24
- NORTH RIVER REVENGE XL
- ACB RESCUE PRO
- METAL SHARK 16 Resolute
- WAR EAGLE
- USCG YARD BUOY BOAT
- ZODIAC H472
- METAL SHARK COURAGEOUS 23
- METAL SHARK DEFIANT 38
- ZODIAC H733 INTERCEPTOR MK-I
- ZODIAC SM557008
- METAL SHARK RELENTLESS 18
- GLADDING-HEARN 65 FT S/EV
- BOSTON WHALER GUARDIAN
- OTHER

SOURCE: USCG data provided to authors, 2018.

This variation in boat types reflects, at least in part, the large number of manufacturers involved. As shown in Figure 5.3, 29 different manufacturers create the boats in the USCG inventory. The top five manufacturers make up only 71 percent of the inventory, and the top ten make up only 90 percent—which means that there are several logistical and maintenance chains involving different parts made by distinct manufacturers. Training requirements are similarly varied.

Figure 5.3. Manufacturers of Boats in the USCG Inventory

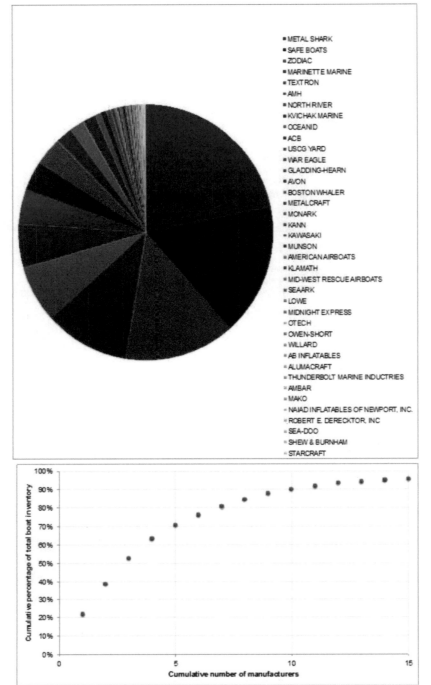

SOURCE: USCG data provided to authors, 2018.

Diversity among boat models from a variety of boat manufacturers also has benefits. Boats performing different missions in different environments, operating either from cutters or from the shore, have distinct requirements. Some manufacturers might fulfill these requirements better than others, and there are obvious benefits to having a healthy marketplace for USCG business. On the other hand, some reduction in the number of boat models being acquired, and in the

43

number of manufacturers making those boats, seems appropriate. We do not have enough data to specify the precise number of boat models and manufacturers required. However, reducing both sets by a factor of two to three would likely reduce risks and costs associated with acquisition, operation, maintenance, and logistics while allowing diverse models of boats and avoiding overdependence on a few manufacturers. Purchasing larger numbers of boats from fewer manufacturers might provide the USCG with more leverage in price negotiations (and enable manufacturers to scale up their operations to reduce unit costs). It is worth noting that NAVSEA is interested in similar standardization across its boat inventory and might be a good resource for shared planning and lessons learned.

Organizational Goals, Logic Models, and Metrics

We recommend that the USCG Boats Acquisition Program continue to develop goals from which verifiable metrics can be derived that are traceable to CG-9 guidance. In determining what funding and organizational structure strategy would be most effective, we needed to determine the overall objectives of the Boats Acquisition office. Through interviews, research, and CG-9325 guidance documentation, we assessed potential funding and structure options based on the impact of the program's ability to

- add additional projects to the office
- be flexible in managing projects
- interface with other acquisition stakeholders
- staff to meet acquisition demands.

We determined these objectives to be the most appropriate based on our research and with the stated goals of the Boats Acquisition Program. Identifying the manner in which program strategies are assessed would be best done in the context of a clearly defined program strategy. CG-9325 is in the process of developing program goals and objectives for such a purpose, which will be linked to the CG-9 strategy. The USCG Acquisition Directorate Strategic Plan provides clear traceability from a vision, to a mission, to goals, to objectives, and finally to performance measures for each objective (with a target value). This type of approach provides a good model to follow in terms of both alignment and format. The CG-9325 goals ought to align with those of CG-9, but the format itself is very clear and traceable.

In particular, one of the stated CG-9 objectives is to "[d]eploy consistent knowledge management tools and documented, repeatable processes."[4] One of the performance measures for that measure is to comply with CG-9 policy for capturing and recording lessons learned. During interviews, we asked if there are any lessons learned that document previous programmatic or acquisition challenges to assist in future understanding of how similar challenges had been solved. We found that, in general, lessons learned are not documented and archived, at least not for the challenges immediately related to the purpose of our study. A

---

[4] USCG Acquisition Directorate Strategic Plan, 2016.

greater emphasis on capturing lessons learned within the USCG might help provide greater historical perspective and possible solutions for some of the challenges the program faces.

Logic Models as a Prerequisite for Metrics

Metrics can neatly capture the extent to which boats acquisition goals are being met and can indicate where challenges exist. A natural first step is identifying which items should be measured; to that end, we developed a "logic model" structure on which that analysis could be based. Figure 5.4 is an example of a basic logic model structure. At the right side of the model are *inputs:* people, funding, authority, physical objects, and infrastructure that are employed to perform *activities*. The direct results of those activities are *outputs*, the direct result of the organization's actions. The *outputs* contribute to the organization's higher-level goals, *outcomes*. Finally, those *outcomes* contribute to *strategic goals* that represent ultimate aims.

Figure 5.4. Structure of a Logic Model

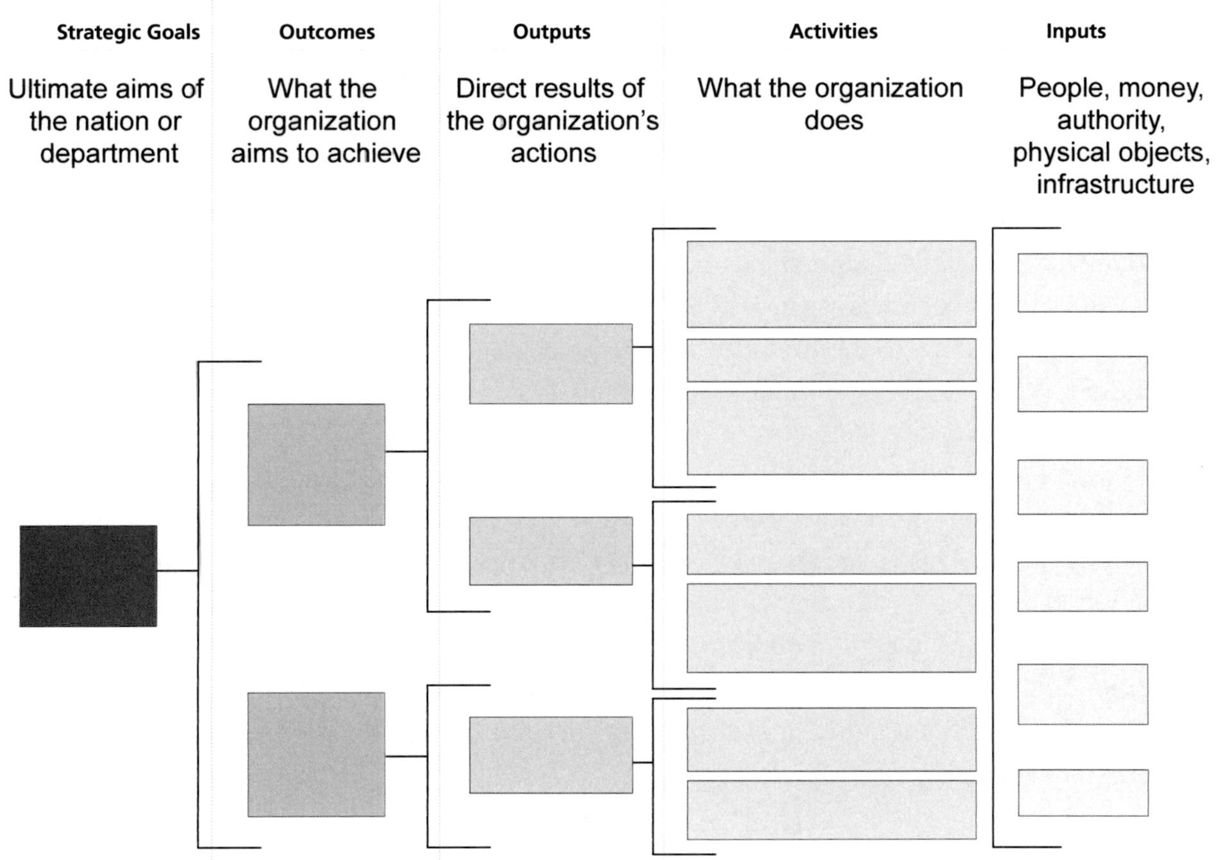

Strategic Goals	Outcomes	Outputs	Activities	Inputs
Ultimate aims of the nation or department	What the organization aims to achieve	Direct results of the organization's actions	What the organization does	People, money, authority, physical objects, infrastructure

We have developed a logic model that characterizes Coast Guard boats acquisition and procurement, as shown in Figure 5.5.

45

Figure 5.5. Proposed Logic Model Describing Coast Guard Boats Acquisition and Procurement

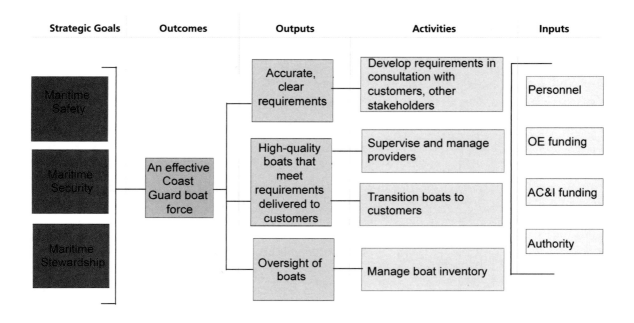

This logic model, like any logic model, is not definitive: There are inevitably opportunities to revise it in order to capture or emphasize specific points, many of which could be brought forth in a series of workshop sessions involving diverse stakeholders.[5] However, the important point is that, to the degree that it characterizes the processes involved, the value of unifying funding streams and offices becomes clear. Although OE funding and AC&I funding come from different sources, both contribute to the funds needed to run acquisition and procurement operations and to spend on boats. Having two funding streams could be an imposition due to political and fiscal reality, but it offers no advantage over a single stream of funding, if that funding can be stripped of rules that limit its fungibility. Likewise, thinking about the activities that constitute procurement and acquisition, it becomes clear that the activities of the former are essentially a subset of the latter. This suggests that having two separate offices handling procurement and acquisition is likely to be disadvantageous: Since many of their activities are the same, as are some of the outputs, it likely makes more sense to combine them. Moreover, the fact that acquisition of a nonstandard boat can later transition into procurement of the same boat, once it becomes a standard commodity, makes the case for uniting both acquisition and procurement within a single office.

---

[5] The development of logic models and metrics is an inherently iterative process, and one that ideally entails teams of personnel collaborating over extended periods. It also must be revisited every few years to reflect changes in the environment and in organizational goals. For more information on this approach, see Scott Savitz, Henry H. Willis, Aaron Davenport, Martina Melliand, William Sasser, Elizabeth Tencza, and Dulani Woods, *Enhancing U.S. Coast Guard Metrics*, Santa Monica, Calif.: RAND Corporation, RR-1173-USCG, 2015.

Identifying Metrics

Identifying metrics for the logic model's inputs is easy: We include numbers of personnel, quantities of money, or amounts of purchasing authority. Metrics for activities are also simple: numbers of person-hours spent on an activity, as well as the percentage of activities that meet previously designated cost or timeline criteria. Output metrics could include the following:

- number of sets of requirements that meet certain thresholds in terms of quality
- number of boats delivered to customers that meet requirements
- number of types of boats delivered to customers that meet requirements while also adhering to schedule and cost constraints
- accuracy and timeliness of oversight of boats.

It is more challenging to characterize metrics for the overall effectiveness of the USCG boat force, which also depends heavily on factors beyond the remit of acquisition and procurement personnel. Two such metrics are

- the percentage of USCG boat-based mission activities (e.g., interdiction, boarding) that are conducted without boat capability limitations hindering operations
- the number of USCG personnel injuries or fatalities per year that can be attributed to shortfalls in boat capabilities or functions.

The development of metrics for USCG strategic goals—namely, maritime safety, security, and stewardship—is beyond the scope of this project. We document them here not for the purpose of identifying metrics but rather to provide a context for the other items in the logic model.

Most of these metrics do reasonably well when evaluated using the following standard criteria:

- **Validity:** the extent to which the metric accurately measures an element of the logic model
- **Reliability:** how consistently measurements can be made
- **Feasibility:** how easily the measurement can be made (the quantity of resources required to make the measurement)
- **Timeliness:** how quickly the measurement can be made.

For our purposes, the important point is that these metrics, which achieve at least a basic level of validity, reliability, feasibility, and timeliness, are applicable to both procurement and acquisition. This points to a further advantage of unifying the two within a single office: Data collection to evaluate these metrics can be conducted in a uniform way across both, facilitating comparisons. Although the acquisition of nonstandard boats is more complex than procurement of standard ones, and therefore the values of particular metrics might be worse for acquisition, having data sets that are being collected and analyzed uniformly can enable comparisons that reveal particular issues.

Future Acquisition Analysis

We recommend that the USCG Boats Acquisition Program conduct predictive analysis on inventory, in coordination with CG-731, to better prepare for new projects. Currently, CG-731 maintains a database to track all USCG boat assets, including the hours on each asset, acquisition dates, and expected service-life limits. Boat stakeholders use the reports generated by this database to understand when boat models will to reach service-life limits, to identify which boats are over- or underutilized, and to make adjustments accordingly. It is important that CG-9325 accesses these data and uses them to help inform acquisition decisions and prepare for future projects. The lead-time for new boat acquisition is typically 36 months; therefore, USCG must understand when boats are roughly within 36 months of their useful life limits to initiate follow-on projects. For example, the top portion of Figure 5.6 shows that the Response Boat–Small inventory ranges between eight and 14 years in age. Although 100 percent of the inventory is within its service-life limit of 15 years, almost 40 percent is within 36 months of its useful-life limit. The lower portion of Figure 5.6 shows the distribution of ages of the entire boat inventory.

Figure 5.6. Boats Inventory Summary

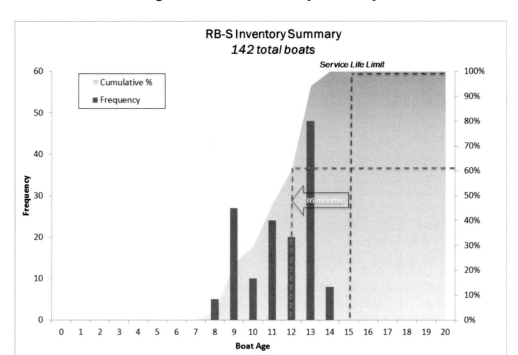

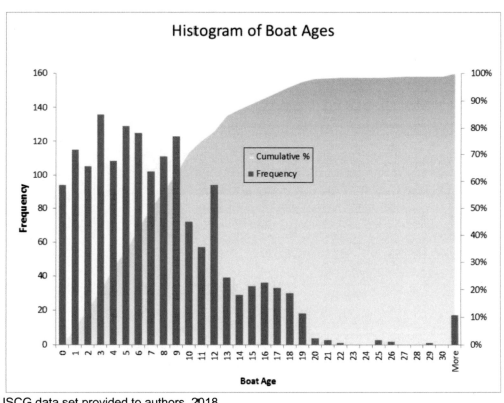

SOURCE: USCG data set provided to authors, 2018.

During the course of this project, we noted that forecast modeling for asset lifecycle could potentially aid in asset lifecycle management, particularly with respect to how and when

49

acquisition decisions should be considered at various levels of the organization. The Small Boat Product Line of the USCG Surface Logistics Center maintains data and measures many factors affecting a small boat's lifecycle. Both CG-731 and CG-9325 collect asset data and expressed interest in being able to develop a more predictive tool to help manage future acquisition efforts. Quantitative and relational analysis of quantitative factors affecting acquisition decision planning might reveal an opportunity for predictive modeling. For example, plans for cutter acquisition or potential changes in demand for particular missions could be used to better anticipate the numbers and models of boats that will likely be needed over the next several years.

Collaboration

We recommend that the USCG explore possible collaboration opportunities with external boats acquisition offices. Some of the external organizations we interviewed expressed an interest in possible collaboration with the USCG on boats projects, especially in cost sharing. Such collaboration could cut costs and create possible maintenance and sustainment efficiencies, although collaborative efforts bring additional challenges (such as meeting requirements for multiple organizations).

Organizations within DHS would be natural USCG partners for boats acquisition. CBP Air and Marine Operations has an inventory of several hundred boats (about one-fifth the size of the USCG's inventory). Immigration and Customs Enforcement also uses boats. Collaborating on boat purchases could provide all three agencies with additional ideas, additional leverage with suppliers, and increased commonality to help promote tactical and operational coordination.

Outside DHS, the Navy, Marine Corps, and Army use boats somewhat differently from the USCG, but their requirements might be similar. In some cases, the collective purchasing power of the USCG and other military services could help reduce costs.

Further afield, there might also be opportunities to collaborate with the Canadian Coast Guard, which is already a partner of the USCG in multiple missions and districts. In a few cases, state or local organizations might be possible partners on requirements or even actual acquisition; police and fire departments across the nation use boats in rivers, lakes, and bays and some tactical aspects of these departments' missions overlap with those of the USCG.

Bibliography

Butt, Mark E., *Cutter Capital Asset Management Plan (CCAMP)*, Commandant Instruction 4700.1, Washington, D.C.: United States Coast Guard, October 9, 2012.

Commandant Instruction Manual 5000.10D, *Major Systems Acquisition Manual (MSAM)*, Washington, D.C.: U.S. Coast Guard, 2015.

Commandant Instruction Manual 5000.11A, *Non-Major Acquisition Process (NMAP) Manual*, Washington, D.C.: U.S. Coast Guard, 2011.

Commandant Instruction Manual 5000.11B, *Non-Major Acquisition Process (NMAP) Manual*, Washington, D.C.: U.S. Coast Guard, 2012.

Commandant Instruction Manual 5000.12, *Management Roles and Responsibilities*, Washington, D.C.: U.S. Coast Guard, 2012.

Commandant Instruction Manual 5400.7F, *Commandant Change Notice 5400 to Coast Guard Organizational Manual*, Washington, D.C.: U.S. Coast Guard, January 27, 2015.

Commandant Instruction Manual 7100-3E, *Financial Resource Management Manual (FRMM)*, Washington, D.C.: U.S. Coast Guard, 2013.

Commandant Instruction Manual 16114.4B, *Boat Management Manual*, Washington, D.C.: U.S. Coast Guard, 2012.

Commandant Instruction Manual 16114.20, *Shore Based Response Boat Strategic Vision and Transition Plan*, Washington, D.C.: U.S. Coast Guard, October 31, 2001.

Department of Homeland Security, *Fiscal Year 2017 Congressional Justification*, Washington, D.C., 2016.

Deyo, Russell C., *Acquisition Management Directive*, Washington, D.C.: Department of Homeland Security, DHS Directive 102-01-R03, July 28, 2015.

Department of Homeland Security, Under Secretary for Management, *Acquisition Management Instruction*, Washington, D.C., DHS Instruction 102-01-001-R01, March 9, 2016.

Ford, Robert C., and W. Alan Randolph, "Cross-Functional Structures: A Review and Integration of Matrix Organization and Project Management," *Journal of Management*, Vol. 18, No. 2, 1992.

Galbraith, Jay R., "Matrix Organization Designs: How to Combine Functional and Project Forms," *Business Horizons*, Vol. 14, No. 1, February 1971.

Kerzner, Harold, *Project Management: A Systems Approach to Planning, Scheduling and Controlling*, 8th ed., Hoboken, N.J.: John Wiley & Sons, Inc., 2003.

Public Law 113-6, Consolidated and Further Continuing Appropriations Act, March 26, 2013.

Savitz, Scott, Henry H. Willis, Aaron Davenport, Martina Melliand, William Sasser, Elizabeth Tencza, and Dulani Woods, *Enhancing U.S. Coast Guard Metrics*, Santa Monica, Calif.: RAND Corporation, RR-1173-USCG, 2015. As of May 2, 2017: https://www.rand.org/pubs/research_reports/RR1173.html

Turner, J. Rodney, and Anne Keegan, "The Versatile Project-Based Organization: Governance and Operational Control," *European Management Journal*, Vol. 17, No. 3, 1999.

U.S. Coast Guard, Acquisition Directorate, *Strategic Plan: Blueprint for Sustained Excellence*, Version 6.1, Summer 2016.